D0060224

BURSTS

Also by Albert-László Barabási

*LINKED: How Everything Is Connected to Everything Else
and What It Means for Business, Science, and Everyday Life*

Albert-László Barabási

BURSTS

The Hidden Pattern
Behind Everything We Do

Dutton

DUTTON

Published by Penguin Group (USA) Inc.

375 Hudson Street, New York, New York 10014, U.S.A.

Penguin Group (Canada), 90 Eglinton Avenue East, Suite 700, Toronto, Ontario M4P 2Y3, Canada (a division of Pearson Penguin Canada Inc.); Penguin Books Ltd, 80 Strand, London WC2R 0RL, England; Penguin Ireland, 25 St Stephen's Green, Dublin 2, Ireland (a division of Penguin Books Ltd); Penguin Group (Australia), 250 Camberwell Road, Camberwell, Victoria 3124, Australia (a division of Pearson Australia Group Pty Ltd); Penguin Books India Pvt Ltd, 11 Community Centre, Panchsheel Park, New Delhi—110 017, India; Penguin Group (NZ), 67 Apollo Drive, Rosedale, North Shore 0632, New Zealand (a division of Pearson New Zealand Ltd); Penguin Books (South Africa) (Pty) Ltd, 24 Sturdee Avenue, Rosebank, Johannesburg 2196, South Africa

Penguin Books Ltd, Registered Offices: 80 Strand, London WC2R 0RL, England

Published by Dutton, a member of Penguin Group (USA) Inc.

First printing, April 2010

10 9 8 7 6 5 4 3 2 1

REGISTERED TRADEMARK—MARCA REGISTRADA

LIBRARY OF CONGRESS CATALOGING-IN-PUBLICATION DATA

has been applied for.

ISBN 978-0-525-95160-5

Printed in the United States of America

Set in Dante

While the author has made every effort to provide accurate telephone numbers and Internet addresses at the time of publication, neither the publisher nor the author assumes any responsibility for errors, or for changes that occur after publication. Further, the publisher does not have any control over and does not assume any responsibility for author or third-party Web sites or their content.

Gyermekeimnek

Contents

1 The Best Bodyguard in the Business   *1*

2 A Pope Is Elected in Rome   *15*

3 The Mystery of Random Motion   *21*

4 Duel in Belgrade   *37*

5 The Future Is Not Yet Searchable   *43*

6 Bloody Prophecy   *55*

7 Prediction or Prophecy   *61*

8 A Crusade at Last   *71*

9 Violence, Random and Otherwise   *77*

10 An Unforeseen Massacre   *91*

11 Deadly Quarrels and Power Laws   *97*

12 The Nagylak Battle   *109*

13 The Origin of Bursts   *117*

14 Accidents Don't Happen to Crucifixes   *129*

15 The Man Who Taught Himself to Swim by Reading   *133*

16 An Investigation   *147*

17 Trailing the Albatross   *155*

18 "Villain!"   *165*

19 The Patterns of Human Mobility   *171*

20 Revolution Now   *183*

21 Predictably Unpredictable   *191*

22 A Diversion in Transylvania   *207*

23 The Truth about LifeLinear   *213*

24 Szekler Against Szekler   *225*

25 Feeling Sick Is Not a Priority   *229*

26 The Final Battles   *243*

27 The Third Ear   *251*

28 Flesh and Blood   *263*

Notes   *271*

Illustrations   *297*

Acknowledgments   *299*

Index   *303*

06/18/08	7-ELEVEN 290 GEORGE STREET NEW BRUNSWICK
06/18/08	USPS 3356870100 86 BAYARD ST NEW BRUNSWICK
06/18/08	NJT NEWARK RAYMOND PLAZA WEST NEWARK NJ
06/18/08	WHITE CASTLE 100060 525 8TH AVE MANHATTAN N
06/18/08	NJT NEW BR...NCH AND ALBAN STREET NJ
06/18/08	7-ELEVEN 290 GEORGE STREET NEW BRUNSWICK
06/17/08	NJT NEWARK RAYMOND PLAZA WEST NEWARK NJ
06/17/08	MTA VENDING MACHINES 130 LIVINGSTON ST 5TH 212-METROCARD NY
06/17/08	DELTA COMMUNITY CCU LAGUARDIA AIRPORT FLUS
06/16/08	KABUL AFGHAN CUISINE SAN CARLOS CA
06/10/08	7-ELEVEN 284 S ELEVENTH ST SAN JOSE CA
06/10/08	THE GARAGE SAN JOSE CA
06/10/08	CALTRAIN TVM SAN CARLOS CA
06/09/08	HAWAIIAN DRIVE INN SAN JOSE CA
06/09/08	CALTRAIN TVM SAN CARLOS CA
06/09/08	SANTA CLARA COUNTY FCU 140 E. SAN FERNAND
06/08/08	Check W/D 376
06/08/08	EL POLLO LOCO #3604 SAN JOSE CA
06/08/08	EL POLLO LOCO #3604 SAN JOSE CA
06/08/08	PEET'S #14302 SAN JOSE CA

Wednesday, June 19, 2002

1

The Best Bodyguard in the Business

Ⅰf all goes well, by the time you close this book I will have convinced you that, despite the spontaneity you may exhibit, you are far more predictable than you are willing to admit. This is by no means personal. I am just as easy to forecast, as is everybody with whom I live and work. In fact, algorithms built in my lab to discover how predictable we are were tested on millions of individuals and failed only once. His name was Hasan. Hasan Elahi, to be precise.

～

It was the anxiety in the air that caught Hasan's attention as he sized up the fifty or so foreigners detained by the Immigration and Naturalization Services at the Detroit Metropolitan Airport. "You could just tell that this was their first day in the U.S., and you could see the fear in everyone," he recalls. "I was just confused—what am I doing here?"

With a passport as thick as a cheap paperback novel, having been extended three times to fit the visas and the border stamps he collected during his travels, Hasan knew a few things about immigration. One of them was that U.S. citizens don't get pulled aside by the INS when they return home. At least, they don't normally. More puzzled than scared, Hasan tried to strike up a conversation with the guards, only to realize that they were just as confused as he was. Finally, a man in a dark gray suit walked up to him and, bypassing any introduction, said in a matter-of-fact tone, "I expected you to be older."

The man was in his fifties, and Hasan thought his greeting extraordinarily awkward, so, in an attempt to break the tension, he tried a light-hearted response. "Sorry, I am trying to age as quickly I can."

It didn't work—the time and place just wasn't going to accommodate humor of any sort. So he decided to get to the point. "Can you explain to me what is going on?"

The man looked at him, paused as if trying to find the right way to put it, and finally just shrugged, replying in a voice devoid of emotion, "Well, you've got some explaining to do yourself."

It was June 19, 2002, and Hasan Elahi, a thirty-year-old media-installation artist, was on his way home after a grueling six-week-long trip that started with a flight from Tampa, Florida, to Detroit, hopping from there to Amsterdam, Lisbon, and Paris, to finally land in Dakar, Senegal. Ten days later he took a forty-eight-hour-long bus ride to Bamako, Mali, and then crossed over to the Ivory Coast. After visiting the largest church in Africa, built to hold three hundred thousand worshippers in a country with only forty thousand Christians, on May 28 Hasan arrived in Abidjan, a major port on the country's southern shore. By then he was tired. "West Africa gets to you after a while. It tests your patience," he recalls, so after the ceiling of his hotel room caved in during a storm, he decided that it was time to get out. He flew back to Dakar, only to find himself on a bus once again a day later, heading to Bissau, the capital of Guinea-Bissau. He then crossed two more borders and got his bleach-blond hair braided with red stripes in Gambia before making his way back to Senegal.

It took Hasan about six more days to finalize his art installation for the Dakar Biennale, after which he returned to Paris, took a train to Strasbourg, walked across the German border to visit a museum famous for its digital art collection in Karlsruhe, dropped by the Documenta exhibit in Kassel, and flew from Hanover to Faro, a tourist resort at Portugal's southern tip. After two days on the beach, he slept at the Lisbon airport the night before his morning flight back to the United States. And so, by this time, smelling quite ripe and with red dreadlocks sprouting out of his head, in a small interrogation room in Detroit, Michigan, Hasan tried to find a reasonable answer to the reasonable question the man in the gray suit had just asked him: "Where were you?"

Where to start? He decided to keep it simple. "Well, I am just coming back from Amsterdam."

"Where were you before that?" came the next question.

"I was in Lisbon."

"Where were you before that?"

"In Faro, on the beach."

Step by step, Hasan traced back the stations of his journey, finally arriving in Dakar.

"Where's that?" asked the man. Hasan looked across the L-shaped wood-grained table and realized that this was not a test or a joke—his interrogator didn't exactly know where Dakar was.

"So, I still don't know what's happening, but I am kind of assuming that this has probably something to do with terrorism and that I am some kind of suspect and this is some kind of government or law-enforcement official I am talking to," he recalls. "Obviously you can't get angry at them, and you can't say that you are an idiot for not knowing what the largest city of West Africa is. As much as you want to say it, you just don't. You have to play cool and act professional."

So with his fingertip Hasan drew on the desk an imaginary map of Africa, pointed to the continent's invisible western corner, and explained Dakar's significance as a transit point to the United States during the slave trade.

"Any Muslims there?" came the next question.

"Yes—about ninety-five percent of the population," responded Hasan, only slightly ironically.

"Whom do you meet at these places?"

"Other artists—you know, people who work in art business, writers, journalists," responded Hasan. He proceeded to patiently explain the day-to-day of the art business.

"What kind of art do you make?"

Once again, this wasn't easy to answer. Hasan is an artist, but not the kind whose work you can hang in your dining room. His pieces are light on aesthetics and dense with ideas, representing witty and, I think, poignant commentaries on our world. Take, for example, the installation he did in Dakar, a fifteen-foot-tall communication tower built from bamboo sticks with a long TV antenna on the top. Four neon lights filled the room with blue light, accompanied by a speaker emitting random hissing sounds. For the uninitiated, none of this makes any sense. Yet each element had its purpose.

The first thing that struck Hasan upon his arrival in Senegal was how blue everything was. "Especially if you were standing at the edge of the ocean, the blue water and the blue sky, it is absolutely gorgeous," he recalls—hence the blue neon lights in his installation. The other thing he noticed was the sharp hissing sound of the Senegalese calling to each other. Inexplicably, a person knew exactly who was hissing at him, even half a block away. So the recording was essential, testing how the Senegalese would react to an artwork hissing at them.

Hasan still laughs as he recalls his interrogator's request to shed light on his art. "I have a hard time explaining this to other artists," he recalls with great amusement, "much less to law-enforcement officials." Given that the work he showed in Senegal was sort of sculptural, he said "I'm a sculptor," and left it at that, certain that mentions of media installations would only confuse things further.

Then, abruptly, came the next inquiry: "Do you have a storage unit near the university?"

"Yeah." Hasan nodded. He had begun renting one after he moved to Tampa to teach at the University of South Florida.

"What do you have in it?"

"Winter clothes I have no use for in Florida, furniture that I can't fit in my tiny apartment, assorted garage-sale material because I am a pack-rat—just, you know, assorted junk."

"No explosives?" the agent asked now, somewhat confused and suspicious.

"No, I am certain I had no explosives in there," Hasan responded.

And with that, question after question, unraveled the reason for his detention. A few weeks earlier the Tampa FBI had received a tip that a man, hoarding explosives in his storage unit, had fled on September 12, 2001. The name of the suspect was Hasan Elahi.

> I can't prove who did it, but I am pretty certain that it was the owners of the storage unit. I knew these people; I would spend hours talking with them. Every month when I went to pay my storage I would sit down to chitchat with them—an elderly couple that had moved from Kentucky to Tampa to start a storage business there.
>
> You need to remember the national psyche. This was the summer of 2002, and the psyche is that if you see something, say something! *Now* is the time to report it, not when it hits the news. They saw a brown guy, and they saw his name, and said to each other, "What kind of name is that? It must be one of those Arab names. He must have explosives!"
>
> This couple, they are not malicious people, they are not mean people, they didn't have anything against me personally. They just simply didn't know any better.

It takes less than ten minutes to realize that, despite his Arab-sounding name, Hasan is not a graduate of an al Qaeda training camp. Born in Bangladesh, he speaks English with a faint New York accent, a holdover from his childhood in Brooklyn, where he lived from the age of seven. Yes, he does have olive skin, but his bleach-blond hair is hardly a telltale sign of jihad. He is the average second-generation kid on the block, who speaks, lives, and feels American. It wasn't long until the FBI agent realized all that and let him catch his flight back to Tampa.

In a normal world, things would have ended just there. But the world was not normal after 9/11, not if you were named Hasan and had brownish skin, even if you were the only person without gunpowder in Florida. So Hasan spent the next five months in and out of the FBI facility in Tampa, where he was questioned for hours at a time.

"Basically, I told them every detail of everything in my life; I did not hold back anything," he recalls without any bitterness in his voice. "You know, the reality is that when you are face to face with those who have the power of life and death over you, sitting across, you do not behave like a rational being. You see yourself doing things, but you do not dare to act on it."

Only five months later, shortly after Thanksgiving, following a grueling lie detector test, one of the FBI agents, a big guy with short-cropped hair, told him that it was all over and he was free to go. Hasan was shocked. *That's it? Just like that, it all ends here and now as if nothing has happened over the past five months?* He looked at the FBI agent and said, "Wait a minute, guys, I'm leaving the country soon. What happens on the way back?"

"Where are you going?"

"Indonesia."

"Oh, you should be careful; there's been a terrorist attack there," replied the agent with a concerned look.

Hasan was taken aback at this bizarre turn. After what he'd gone through, this guy must be joking, he thought. But the agent did look genuinely worried. So he decided to speak his mind. "Look, guys, my biggest fear is not a plane crashing or a building blowing up. My biggest fear is one of *you guys,* thinking in your hearts and your mind that you are doing the right thing, takes me away and no one knows where I am and no one knows how to get me out of this mess."

At a time when the United States had just begun hauling prisoners from all over the world to Guantánamo, Cuba, Hasan could see that the FBI agent got his point and that his concern was real. The agent said nothing, but his body language and demeanor told Hasan, *Yes, things like this do happen these days, and I am worried for you.* Encouraged, Hasan pressed his case. "All we need is the last guy not to get the last memo, and here we go all over again. What do I do now?"

The FBI agent thought about this for a few seconds. Then he reached for his wallet, pulled a card from it, and handed it to Hasan, saying, "Here are some phone numbers, and if you get into trouble, please call us." Then, after a short pause, he added, "We'll take care of it immediately; we'll get to it right away."

Hasan looked at the card, then at the agent. Then, somewhat relieved, he said, "Wonderful." Finally, with a hint of humor, he added, "Wow—I've got the best bodyguards in the business."

~

Most of the technological advances we enjoy today, from computers to cell phones, space travel to new drugs, rest on hundreds of years of scientific inquiry driven by an unwavering belief that natural phenomena can be understood, described, quantified, predicted, and eventually controlled. The benefits of this conviction, bordering on the obsessive for many of us in the sciences, are everywhere around us. We learned to control the flow of electrons in semiconductors to build transistors and iPods; we deciphered the laws governing radio waves to talk wirelessly via cell phones; we understood the role of chemicals in our body to provide cures for common diseases; we discovered the laws of gravity, offering us a ticket to the moon.

Unfortunately, this enlightening revolution came to a halt at the outer gates of the natural sciences, never reaching an area that is, nevertheless, under ever increasing scrutiny today: the behavior of individuals and human societies. When it comes to our fellow humans' actions, the sequence of events we witness on a daily basis appear to be just as mysterious and confusing as the motion of the stars seemed in the fifteenth century. At other times, though we are free to make our own decisions, much of our life seems to be on autopilot. Our society goes from times of plenty to times of want, from war to peace and back to war again. It makes you wonder, do humans follow hidden laws, laws other than those of their own making?

Hasan's story is a case in point: Was his encounter with the FBI an accident or something foreseeable given his skin color, name, and, most important, behavior? Does his experience fit into the collection of rules and outcomes that are acceptable in a society like ours? Were the couple from Kentucky just doing what they were supposed to do, fitting neatly into a complex web of patriotism and fear that characterized the world's post-9/11 psyche? Are our actions governed by rules and mechanisms that might, in their simplicity, match the predictive power of Newton's law of gravitation? Heaven forbid, *might we go as far as to predict human behavior?*

Until recently we had only one answer to each of these questions: We don't know. As a result, today we know more about Jupiter than the guy who lives next door to us. We can predict where an electron will go, we can turn a gene on or off, and we can even send a robot to Mars, but we are lost if asked to explain or predict phenomena we might expect to know the most about, the actions of our fellow humans.

The reason is simple: In the past we had neither the data nor the tools to explore what we really do. Indeed, bacteria don't get annoyed at you when you put them under a microscope, and the moon will not sue you for landing a spacecraft on its face. Yet, none of us wants to submit to the invasive inquiry to which we subject bacteria and our planets—aiming to know *everything* about them, *all the time.*

Well, except one person.

❧

Inspired by his new sense of security, Hasan settled on a new routine. Each time he planned a trip abroad he dialed the number on the card the FBI agent had given him and shared his travel plans. "It wasn't that I had to call him. I chose to call him. I wanted him to know this is where I was going, this is what I was doing," he explains. "I am not making any sudden movements; I don't want to raise any red flags."

The calls turned into e-mails, and with time he started to send photos from his trips and brief accounts of his experiences. Gradually the big man with the short-cropped hair stopped being simply a law-enforcement official to Hasan. He began to think of him as *my FBI agent.* And his "bodyguard" lived up to his promise—Hasan came and went without ever being harassed again.

Then a year and a half after Detroit, in January of 2004, after having sent dozens of itineraries and hundreds of photos to the FBI, Hasan had a revelation. Why was he sharing this information only with the FBI? Why not with everyone?

"What if someone makes a mistake, what if something goes wrong?" he asked himself. "These guys know lot about me, but how thorough is their data? They must have missed something about me!

"So that's when I started to create a parallel database of my own, trying to re-create my FBI file. Not just to re-create it, but rebuild it to a much higher level of precision than they had."

He began uploading each photo, together with his current coordinates, to a Web site he created. The routine quickly turned into an obsession that continues even today. Indeed, a visit to www.trackingtransience. net reveals a flashing red arrow on a map, indicating Hasan's current whereabouts. The image above it provides a glimpse of his surroundings, perhaps a picture of a hotel room, a coffeehouse, or an airport. Under the map a series of icons serves as a gateway to a collection of about thirty thousand images and countless factoids regarding his past locations, from photos of the meals he consumed to the urinals he wetted; from a complete list of flight numbers he took to a thorough accounting of his expenses.

By making all his data public, he flipped surveillance upside down, turning the observed into the observer. He was now the suspect *and* the FBI, following himself on their behalf. With that the basic notions of privacy ceased to apply to him, turning him into a unique specimen about whom we can know just about everything. His life became his biggest art project, with a never-ending quality to it, bits of it now being displayed in museums, exhibit halls, and galleries all over the world.

There is an accidental dimension to his art; however, there is something he could not have intuited when he walked down this path. By recording and sharing his activities and whereabouts, Hasan amassed real-time information of incredible detail on one of the most neglected areas of scientific inquiry, which is, ironically, himself, Hasan Elahi.

∼

Hasan is not completely alone. A decade ago Gordon Bell, principal researcher at Microsoft Research, started wearing a digital camera that automatically snapped a picture of each person in front of him and a recorder that captured a wide spectrum of the sounds around him. He also saves a digital fingerprint of just about everything he touches on- and off-line, contributing to an archive that over the past decade has swollen to more than a hundred thousand e-mails, tens of thousands of photographs, the recordings of all the phone calls he has made, close to a thousand pages of health records, every book from his library, and even images of the labels of wine he has tasted.

Or consider Deb Roy, a computer scientist at MIT's Media Lab, who, before his son's birth, installed eleven video cameras and fourteen microphones in his house. The recordings are streamed to the basement and saved there on a terabyte-capacity disc array. Since the system was installed each cry or giggle of his baby, every diaper change, each chat or spat between him and his wife, is stored in more than 250,000 hours of video.

For the rest of us, without the obsessive commitment or the resources of Elahi, Bell, or Roy, there are larger forces at work: a covert but increasingly detailed surveillance we are all subject to. Indeed, today just about everything we do leaves digital breadcrumbs in some database. Our e-mails are preserved in the log files of our e-mail provider; time-stamped information about our phone conversations sits in our phone company's vast hard drives; when, where, and what we shop for, our taste, and our ability to pay for it is cataloged by our credit card provider; all our Web pages, MySpace and Facebook profiles, and blogs are saved and cross-listed in multiple servers; our momentary whereabouts are available to our mobile phone provider; our face and fashion is remembered by countless surveillance cameras installed everywhere, from shopping malls to street corners. While we often choose not to think about it, the truth is that our life, with minute resolution, can be pieced together from these mushrooming databases.

To be sure, the very existence of these records raises huge issues of privacy, a problem whose magnitude cannot be overstated. It also creates a historic opportunity, however, offering for the first time unbiased data of unparalleled detail on the behavior of not one, but millions of individuals. In the past few years these databases have found their way into a variety of research labs, allowing computer scientists, physicists, mathematicians, sociologists, psychologists, and economists to pore over them with the help of powerful computers and a wide array of novel technologies. Their conclusions are breathtaking; they provide convincing evidence that most of our actions are driven by laws, patterns, and mechanisms that in reproducibility and predictive power rival those encountered in the natural sciences. These findings are not constrained to the scientists' sandbox; some of the patterns and laws are already worth billions of dollars, as illustrated by the market capitalization of Google and Yahoo! and other companies whose business model is built on mapping human behavior. They also turned the world upside down. In the past, if you wanted to understand what humans do and why they do it, you became a card-carrying psychologist. Today you may want to obtain a degree in computer science first.

With that we have arrived at the fundamental goal of this book: I will show how our nakedness in the face of increasingly penetrating digital technologies creates an immense research laboratory that, in size, complexity, and detail, surpasses everything that science has encountered before. By following the trail of these discoveries we will come to see the rhythms of life as evidence of a deeper order in human behavior, one that can be explored, predicted, and no doubt exploited. The insights to be gleaned require us to stop viewing our actions as discrete, random, isolated events. Instead they seem to be part of a magic web of dependencies, a story within a web of stories, displaying order where we suspected none and randomness where we least expect it. The closer we look at them, the more obvious it will become that human actions follow simple, reproducible patterns governed by wide-reaching laws. Forget dice rolling or boxes of chocolates as metaphors for life. Think of yourself as a dreaming robot on autopilot, and you'll be much closer to the truth.

❧

Despite the facts of my first story, this book is ultimately neither about Hasan nor the Department of Homeland Security. My goal instead is to address what is normal and what is unique when it comes to human activity. That, however, will take us back to Hasan and his subsequent encounters with the FBI. We will see that the protection offered by his agent worked for a while but failed once the government stopped relying on zealous citizens to track down malefactors. To understand why Hasan was caught once again in the unforgiving web the U.S. government has thrown around the globe in the name of antiterrorism, we will let our computer pore over the data collected on his Web site and compare his behavior with that of millions of other people around the world. The astonishing conclusion? Somehow Homeland Security got it right—at least in discerning that Hasan's behavior is anything but normal. He is a true outlier, important to us precisely because, when it comes to his actions, many of the regular patterns of human behavior we uncover in this book do not apply.

Hasan, however, is not the only outlier in this book. Starting in the next chapter, we will jump back to the time of Luther, Copernicus, and Michelangelo, when astrology, miracles, witchcraft, ghosts, fairies, and omens seamlessly blended with life. We do so to trail a sequence of events that are as fascinating as Hasan's contemporary journey. There we will eventually meet another outlier, a man known as György Székely to his contemporaries. Székely is not the name he used in his hometown. Like many men who roam the world, he was known by the name of the nation he stemmed from, the elusive Hungarian-speaking tribe of the Szeklers, who see themselves as descendants of Attila the Hun. They live in the breathtaking Carpathian Mountains of eastern Transylvania, where Bram Stoker placed Count Dracula and where I, too, happened to be born. As we proceed with our journey into science and history, we will see that, just like Hasan, György Székely is in many ways rather ordinary. Yet, thanks to his unpredictable reactions to the often-strange turns of history, he ends up leading a papal Crusade to victory, without ever getting anywhere near his enemy.

Yet it is not merely György Székely's outlier status that interests us. Rather it is the fact that just about everything he did while forcefully changing history was foreseen by one of his contemporaries. That is, he appears to have been, just like you and me and everybody else around us—well, everyone except for Hasan—deeply predictable. Still, it is one thing to predict your whereabouts today, relying on all-penetrating technologies that monitor all of us; it is a wholly different matter to foresee battles in the sixteenth century and the actions of popes, cardinals, and warriors. How could anyone back in the sixteenth century so accurately foresee his nation's destiny?

Mark Twain said once that history does not repeat itself, but it does rhyme. Let's listen to this rhyme, allowing the obvious differences between Hasan Elahi, György Székely, and you and me and millions around us to guide us toward the deeper parallels behind our actions. Indeed, no matter how fascinating some of the minute details of life are, science shines in its ability to uncover the generic and the universal. Regarding our behavior as humans, this is our pursuit, a glimpse of the universal.

2

A Pope Is Elected in Rome

The Vatican, March 10, 1513, a year before the Papal Crusade

illing suspension of disbelief, to paraphrase Coleridge, is required for us to proceed. We are about to travel back to the early sixteenth century and become a fly on the walls of the freshly painted Sistine Chapel. It is a good vantage point from which to witness a pivotal moment of history, whose real impact will play out hundreds of miles away, in the distant kingdom of Hungary. As we observe the unfolding of these events, I will anchor the story as much as possible on facts. Of necessity, for lack of the cameras omnipresent in the Sistine today, we will let our imagination fill in the details. As you may question, rightly so, whether it really happened *this* way, keep in mind that imagination is at the heart of all innovation. Crush or constrain it and the fun will vanish. So, bear with me, suspend your disbelief, and let the action guide your intellect and instincts.

✿

The misery of the past five days following the pope's death had taken its toll on Cardinal Bakócz. He had been forced to leave his elegant villa in favor of temporary residence in a dark, tiny cell hastily built inside the Sistine Chapel. Yet, while his body was frail, he understood the significance of the upcoming ballot. The haggling was over, and he welcomed the vote. He was more than ready for the election of the new pope.

He paused for a second to discreetly observe the cardinals around him, and his eyes saw only privilege. Many of them were still children, as far as Bakócz was concerned. Cardinal Petrucci, only twenty-two. Cardinal Cornaro, barely thirty. Giovanni de' Medici, thirty-seven. There is no way they could have *earned* the cardinal's zucchetto at that tender age.

Giovanni de' Medici, whom he himself had just negotiated back to freedom from his French captors, was a case in point. This Medici had been made the abbot of Font Douce before he could even speak and was promoted to cardinal by the age of thirteen. This had been the late pope's gift to the Florentine clan for marrying Giovanni's sister to His Holiness's bastard son. Yet even the pope feared that he had gone too far, thereafter denying the child the privileges of the cardinal's office for three years.

To be sure, the powerful Medici clan would have loved to see their Giovanni seated on St. Peter's throne, but since no one in the chapel was disposed to wait decades for the next vacancy at the Holy See, for once the youth was not necessarily an asset. An old pope, like Bakócz, guaranteed not to last, was the best they could hope for.

As the cardinal forced his attention back to the ballot before him, the smell of Michelangelo's paint, still drying on the chapel's ceiling, surprised him with a hint of a memory. He paused once again, distracted by the vision of the wheelwright's house the scent had summoned. It was the odor of the place he had once called home. Unlike Medici and the many others in the chapel, he had never been granted any privilege. He was raised in poverty by the benevolent hand of providence, as he was fond of reminding his surroundings occasionally. The smell of the paint now reminded him how low he had started and how far he had risen.

The cardinal looked again at the ballot in front of him, keenly aware that this round of voting was only intended to clear the field. It was no secret to anyone who was wanted by the people of Rome and those of influence. Five days ago, when he celebrated the Mass of the Holy Ghost, the last public event before the conclave, the field of serious contenders had already been whittled down to three: the Italian cardinal Riario, the Venetian cardinal Grimaldi, and himself, Cardinal Bakócz of Hungary.

Was he delusional in thinking that he had a chance? Perhaps, but he had not counted on benevolent providence alone; one by one, he had collected his votes. The first promise of support had come eight years before from Emperor Maximilian. He had behind him the Hungarian court as well, where he was already "the pope, the king, everything he ever wants to become." And should the candidacy of the Venetian Grimaldi falter, Bakócz could count on the doge to deliver the Venetian votes. At the same time, twelve of the twenty-five cardinals were Italian and had made it clear that they wanted one of their compatriots for the job. Once this tangled web of loyalties had been sorted out, the field was clear: The candidate to beat was the Italian cardinal Riario, nephew of the late pope Sixtus IV.

The cardinal wearily touched his pen into the ink and, ignoring the screeching sound of his quill, wrote a name on the ballot in front of him. He then folded the paper lengthwise and held it aloft for everyone to see, proceeded to the altar, and slowly dropped it onto the paten. He lifted the paten to the full view of his fellow cardinals and slowly slid the ballot into the chalice.

While the past five days had positively flown by, the mere minutes it took for each of the twenty-five cardinals to ceremoniously proceed to the altar, one by one, to place his ballot into the chalice, felt like an eternity. Finally the procession was over, and the young but frail Giovanni de' Medici, the cardinal-deacon, started to read the ballots with soothing modesty. He had undergone surgery only the day before and yet performed his duty diligently.

As he listened to Medici's voice, the cardinal laid his hands on his soutane to hide the slight tremble that slowly conquered his left, the toll of emotion, stress, and seventy-two years of living. With a strong will to not give up right now, he forced himself to follow the votes bouncing on the Sistine's walls.

"Cardinal Serra," Medici read the first ballot, placing it carefully in a separate pile.

"Cardinal Bakócz," he read the next one, pushing the cardinal's heart into an intense dance. It was the only time Bakócz had ever stepped inside the conclave, and yet here he was, among the *papabili*. While it was too early to declare victory, his soul ached for it, overriding his brain's cautionary signals.

His subconscious got another boost when Medici looked directly at him as he announced the next vote: "Cardinal Bakócz."

Two votes for him, two of the first three. A pope needs a two-thirds majority to ascend. And while that rarely happened at the first scrutiny, so far Bakócz had it.

Cardinal Bakócz. Cardinal Rovere. Cardinal Serra. Cardinal Finale. Cardinal Bakócz. Cardinal Serra. Cardinal Serra. Cardinal Grimaldi. Cardinal Serra. Then, hesitating for a second, the cardinal-deacon read his own name: "Cardinal Medici." After this deliberate show of modesty, he continued to calmly open the remaining ballots.

As the Bakócz tallied the numbers in his head, a mix of deep emotion and humility supplanted his fatigue. Serra got most of the votes, fourteen, but remained short of the required two-thirds majority. The elderly, Spanish Serra was not truly papal material, so many cardinals, keen to hide their real intentions, voted for him.

What mattered, they all knew, were the three *papabili*. Of those, the Venetian Grimaldi got only two votes. That was excellent news for Bakócz, as with Grimaldi out he could now count on the Venetian vote during the next scrutiny. Rafello Riario, the hope of the Italian cardinals, had received not one vote, a significant victory. At the end of the first round, Bakócz was certain that it had escaped no one's attention that first among the *papabili* was himself, Cardinal Bakócz.

He could not remember how many decades had passed since he had stopped believing in miracles. Yet, with a humility that he had not felt for years, he looked at Michelangelo's God, grateful to have experienced one before he died. Most visitors were drawn to God's gentle touch as he sparked life into Adam. But lately Bakócz had befriended a different God, the one depicted two panels away, the forceful creator who commanded the sun and the moon into motion.

What had happened in the chapel today had been more than a mere spark; it had been akin to the orchestration of the fluid but relentless motion of the planets. The way had been prepared by fourteen years of lobbying and money sufficient to enrich two kingdoms. Twenty years of friendship with Venice had earned him the patriarchy of Constantinople, an office second only to the papacy. And his triumphant entrance into Rome a year ago, with a pomp that overshadowed that of the pontiff himself, gave him the popular vote.

Granted, he had opened wide many doors with gold and persuasion, some of which he had found only briefly cracked. Regardless, he had quietly squeezed through. He had blown gently on the rolling dice, over and over, until his name ended up on the top.

With all other credible contenders voted out of the race, there remained only one vote that stood between him and the papacy. Bakócz raised his eyes to the other cardinals, offering them an opportunity to signal submission to their future pontiff. Out of the corner of his eye he noticed the loser, Cardinal Riario, slowly making his way toward the ill Medici's private quarters. Touted as the top contender, he had ignominiously failed to win a single vote and surely needed the solace of his ailing friend.

Cardinal Bakócz then noticed Adriano Castellesi, Medici's zealous opponent, follow Riario into the chambers, and his stomach lurched. It was not the mixture of impatience and fatigue he had felt earlier, while waiting for the scrutiny's outcome. This spasm came deep from his gut, a sensation that strong and aggressive men like him rarely experience. It was the naked fear the early Christians must have felt when they saw the lions rushing through the gates toward them, realizing, in the midst of the thundering crowd, that no matter how strong their faith, they were to be eaten alive.

As Bakócz watched the two men, draped in long ceremonial gowns, gravely follow Medici's sedan chair, he knew even before his mind could put it all together that something had gone terribly wrong.

The Mystery of Random Motion

Before 2002 the only dollar bill that could spark the interest of Gary Kanis was the one he could walk away with. On January 12, 2002, sitting at his table covered with neatly arranged boxes of Remington and Winchester pistol bullets, .44 Magnum nickel brass and flat-point hard-cast lead, he had no way of knowing that the end of his one-dimensional relationship with money was nigh. Soon the disposal of record numbers of greenbacks was about to become the greatest satisfaction of his life.

His table was only one of the several hundred that filled the huge hangar at the Eastwood Expo Center Gun Show in Niles, Ohio. Most stalls offered random collections of guns—from the 1920s DWM Luger German automatic pistols to the Remington Model 700 Police rifle, bolt-action sniper rifles with adjustable bipod. Others catered to the twisted fantasies of the gun subculture: military uniforms, World War I helmets, knives, and Nazi insignia. It was a cacophony of male pride gone weird.

Were you curious about the values of Gary's customers, a quick survey of the extensive bumper sticker collection at a nearby booth would clear up any confusion. For three dollars you could walk away with a prominent yellow sign that bellowed, "Welcome to America. Now either speak English, or leave." For another three bucks you could take on the other half of the country: "A woman's place is in the house. But not in the White House."

In his late fifties, with a silver-tipped mustache, neatly parted and combed salt-and-pepper hair, and a slight furrow in his brow, Gary isn't your stereotypical gun dealer. A bit paunchy and not overly scrupulous of his appearance, he yet retains a tidy look and demeanor. He used to own a brass shop back in his hometown of Wattsburg, Pennsylvania, about fifteen miles from the shore of Lake Erie in the northwest corner of the state. But in Wattsburg, population 378, 17 percent of the families lived below the poverty line, and per capita yearly income rarely topped $14,000—dismal conditions for operating a profitable business, even one selling guns. No, the townsfolk couldn't balance the ledgers, and so Gary travels, setting up his booth at about twenty gun shows each year, hauling with him boxes of brass he purchases from wholesalers and a thick wad of cash from which he makes change. A modest Web site ostensibly extends his business; for a reasonable fee he is happy to ship your brass order to you via UPS. But there isn't much demand for the service—most of his customers prefer to pay cash and pick up their merchandise in person. So gun shows, like this one in Niles, Ohio, where Gary personally delivers the ammo, are the lifeline for his small business.

On January 12 things were running their normal course. Then a customer picked up a box of brass from Gary's table and handed him a ten-dollar bill. Returning the change, Gary noticed something odd on the bill—somebody had stamped WHERESGEORGE.COM in bright red ink onto the face of the currency. Intrigued, Gary set the bill aside for closer scrutiny.

It wasn't until three days later, sitting at home before his personal computer, that he had the time to type the address into the browser. When he pressed RETURN and the page loaded, George Washington's eye stared back at him, and a banner in the familiar typeface of the U.S. Treasury identified the site as THE UNITED STATES CURRENCY-TRACKING PROJECT.

Browsing WheresGeorge.com, Gary quickly realized that the world he had entered with the click of a button was quite different from the one he had left behind in Niles. Gun shows attract anywhere from five to seven thousand visitors, most of whom appreciate discretion. I once asked Gary if he would mind sending me a picture of himself and his wares at a gun show. He promptly answered no. Quick to remove any offense, he explained that it was not because he did not want to share, but after twenty-five years in the business he does not have a single picture.

"If you are found carrying a camera at the show, you are escorted to the door," he had e-mailed me, adding that "the best way to get an idea of a gun show is to attend one." So I did. And I hadn't even entered before I understood Gary's caution. A big yellow sign at the entrance spelled out the rules:

GUN SHOW
Under 21 with parent only
NO loaded guns, clips, or mags
ALL guns must be tied
Open action, have your
gun ready for inspection
NO cameras or recording devices

The insular, paranoid, fiercely private world of gun sales was in stark contrast to WheresGeorge.com, an open doorway to where our money goes. It is similar to Hasan Elahi's not-so-private universe—lots of cash, but no secrets or hidden tricks; every move is tracked and in public. And that is, of course, exactly the point.

~

In March 2004 Dirk Brockmann flew to Montreal to attend the annual meeting of the American Physical Society. About seven thousand researchers and students annually descend on a major American metropolis for the APS's March meeting, but this year they traveled to Canada. With more than six thousand talks spread over five days, the meeting is an exhausting marathon for speakers and attendees alike.

In his mid-thirties, with a clean-shaved head that lends him a Mephistophelian appearance, Dirk was in 2004 a junior physicist at the Max Planck Institute for Dynamics and Self-Organization in Germany. He is soft-spoken, and his occasional, ponderous midsentence pauses give his speech a contemplative quality.

"It's a personal opinion, but I don't like these large meetings a lot," he told me once, allowing a lull to punctuate his thoughts. "I travel ten thousand miles, and I listen to all these talks, and maybe two actually leave a trace in my mind. It is amazingly inefficient, you know." Nevertheless, here he was at the conference, having given his talk, prepared to listen to scores of other lectures, gamely ready to reunite with his colleagues and friends for working breakfasts and happy hours and dinners. After five days, exhausted by the high density of physicists, he left Montreal to visit his former college buddy, Dennis Derryberry.

Dennis was the first student Dirk met as an undergraduate at Duke University, and they instantly hit it off. Dirk remembers his friend as one of those witty and intelligent folks whose excellence seems effortless and whose lack of ambition borders on criminal indolence. Financially independent after a string of jobs as poet, songwriter, and musician, Dennis settled with his family into a cozy log cabin situated in the verdant forests of Vermont, where he earned his living as a cabinetmaker.

One evening during Dirk's visit, the two men found a seat on the cabin's porch and, bundled against the cold mountain air, sipped at even colder bottles of beer. Dennis asked his buddy, "So, what are you studying these days?"

Dirk's interests had lately turned to a problem that at a first glance may have seemed irrelevant to his training as a physicist: He wanted to know how infectious diseases spread. As recent headlines about swine flu or SARS remind us, in a globalized world viruses become an increasing threat. Thirty to 60 percent of the entire population of Europe had been wiped out in the fourteenth century by the Black Death, which had swept like a wave across the map. Over the course of a few decades, by foot or on horseback, the disease made its way across the continent, conquering village after village. In comparison, today a virus can infect a person in Mexico City or Hong Kong, be carried by its host by airplane to Toronto, and in a matter of a few *days* Canadian disease-control officials are scrambling to deal with their own outbreak. The question about the next deadly pandemic is not *if* it will happen—but *when*. And, once it is here, how many people will be affected?

How to halt the next outbreak is not a question of biology and virology, as vaccines against new strains take months or even years to develop, by which time there might be no one to cure. The best short-term defense is to prevent the spread of the virus. And to do this we must first understand how people move.

So, in answer to Dennis's question, Dirk judged that discussion of the apocalypse was ill suited to the idyllic Vermont evening and simply answered, "I want to know how people travel—how often they travel a certain distance, for example."

Dennis absorbed Dirk's answer and then asked, "Have you ever heard of the Web site WheresGeorge.com?"

Dirk had been in Germany for the past fifteen years and had not come across it. And so, the next morning Dennis obligingly showed him the site. Dirk instantly knew that he had found a gold mine.

ᴘᴇ

WheresGeorge.com tracks the path of each dollar bill spent, supporting an absorbing hobby for thousands around the United States. The site works like this: You type in the serial number of a United States denomination with your ZIP code, effectively informing the site of the bill's current location. Next you write or stamp WHERESGEORGE.COM onto the bill, and spend it as usual. Anyone noticing the Web address might be compelled by curiosity to visit the site, where he would enter the bill's serial number along with his ZIP code, recording the bill's new location. The thrill for the finder is to discover the bill's history, as the site will display all of the bill's previously recorded sightings on a map of the United States. The farther the bill goes and the more hits it gets, the greater the bragging rights of the person who first registered it.

The record holder is a bill stamped—or "Georged," to use the lingo of veteran Georgers—on March 15, 2002, in Dayton, Ohio. Two months later it reemerged about 229 miles away in Scottsville, Kentucky, and a month later it was spotted twice, first in Chapel Hill and then in Unionville, Tennessee. It then vanished for half a year, reemerging in Milton, Florida. During the next five months it was registered several times in Texas before it jumped to Panguitch, Utah. Its last sighting was on March 26, 2005, in Rudyard, Michigan, more than three years after its release. All told, the bill traveled 4,191 miles at an average speed of 3.8 miles a day, the pace of a comfortably walking adult.

In the months following his first encounter with the Georgers during the gun show, Gary Kanis kept noticing bills stamped with WHERES GEORGE.COM. He found one in February and typed its serial number into the system before spending it. He found two more bills in March, another two in April, and one in November, dutifully registering each of them. By mid-December he took a closer look at the Web site, and as he puts it with a rueful laugh, "It's been downhill ever since." Between December 10 and the end of the month he registered 1,024 bills, mostly singles, and spent them all.

In hindsight, Gary's beginning was modest. Indeed, six years into his hobby Gary has stamped over 1.1 million bills, altogether worth more than $3.5 million, an average of 340 bills every day—weekdays, weekends, and holidays. In dutifully recording the mobility of each bill, Gary has become the top player in a game that, unbeknownst to him, allows us to open a window of unprecedented detail on human activity.

❧

Midway through 1905, the year often referred to as his *annus mirabilis*, a twenty-six-year-old Albert Einstein scratched a letter to his best friend, Conrad Hebrich. Ostensibly lighthearted, Einstein addressed Hebrich as "frozen whale, you smoked, dried, canned piece of soul." Yet under the teasing is a note of urging, as he presses Hebrich to finish his long-overdue doctoral thesis. As an incentive, Einstein offers that upon the receipt of the dissertation he, too, will share five of his manuscripts under preparation.

He calls the first of the five "very revolutionary," as it "deals with radiation and the energy properties of light." Indeed, fourteen years later the paper would win him a Nobel Prize. The second focuses "on the determination of the true size of the atoms" and is today Einstein's most-cited publication. But Einstein's transcendent fame, the renown that makes him a household name, is owing to the fourth manuscript, which, he writes his friend, is "only a rough draft at this point." Once finished, it will be known as the theory of relativity.

It is the third paper, however, that is of interest to us. In 1905 Einstein promised his friend that it would focus on the "unexplained motion" of tiny objects dropped in liquids, as reported in 1828 by Robert Brown, a British botanist, who had noticed that pollen that fell in dew drops underwent jittery, irregular motion. What is particularly strange about Brown's observation is that the pollen never stopped moving. He soon observed that dust suspended in water moved in the same jerky fashion as the pollen, eliminating the possibility that the pollen particles moved because they were alive. That left him and his contemporaries with a puzzling question: What is this mysterious *perpetuum mobile* that keeps the pollen moving?

A possible explanation called for the existence of tiny, fast-moving water "atoms" that kick the pollen in random directions. Imagine a large balloon placed in the middle of an excited crowd. As the individuals move around, they push the balloon from all directions. As a result, sometimes the balloon will move to the left, occasionally to the right, overall displaying a random, jittery motion. This image is quite appropriate: A particle of pollen is about 250,000 times larger than a water molecule and behaves like a really huge balloon in the midst of a dense crowd.

There is only one problem with this explanation: In 1905 the physical reality of atoms and molecules was not a foregone conclusion. Wilhelm Ostwald, one of the most influential physicists of the time, believed that atoms were only an illusion, while Ernst Mach, Einstein's idol and later nemesis, was skeptical of anything that could not be directly seen.

Einstein, with little respect for authority, shrugged off the critics and set out to find evidence that "would guarantee as much as possible the existence of the atoms of definite size." He started by asking a simple question: If we accept that it is indeed the atoms that kick the pollen in random and unpredictable directions, how far would the pollen travel with time?

The question makes little sense at first: If the pollen moves in an unpredictable manner, it is obviously impossible to predict its future whereabouts. Einstein did not dispute that. He realized, however, that he could still calculate a number of characteristics of the pollen's trajectory. Using relatively simple math and a healthy dose of intuition, he predicted that the particle typically moves a distance proportional to the square root of the time elapsed since its release. That is, if you wait four times longer, given its haphazard back-and-forth motion, a particle will have drifted not four times farther, but only twice the distance from its release point.

This was, however, only a prediction and could not be confirmed without a record of the pollen's trajectory. The data became available three years later, when Jean-Baptiste Perrin, a French physicist, developed a technique to track the motion of small particles suspended in liquids. The agreement between the experiment and Einstein's predictions settled the century-long dispute over the existence of atoms. It also landed Perrin a Nobel Prize in 1926. If Einstein had published only the calculation on the random motion of atoms in 1905, Perrin would probably have shared the prize. But five years earlier he had received a Nobel for the first paper of his 1905 output, so he had to sit this one out.

❧

Humans are not that different from pollen suspended in water. Driven for reasons that are just as enigmatic as the pollen's motion, we display a ceaseless desire to move most of the time. We are not kicked by tiny, invisible atoms but dragged by the imperceptible flickering of our neurons, which we translate into tasks, responsibilities, and motivations. I have no doubt that, should we retrace your trajectory, you could recount in detail each trip you took yesterday or even two weeks ago. Yet to those unfamiliar with your daily routine and responsibilities, your trajectory might look just as unpredictable as did the pollen particle's zigzag motion under Brown's primitive microscope.

While it is nobody's business when and where we go, as Dirk explained to his buddy in Vermont, our mobility is the main reason infectious diseases threaten the world. If I am infected by a virus, as soon as I leave my house I run the risk of passing it on to anybody I meet. Therefore, to predict the course of an epidemic we must first understand where people are and where they take their diseases. Since human motion appears to be just as unpredictable as the pollen particle's trajectory, it is fair to assume that we move randomly. Therefore, Einstein's 1905 ideas about the random atomic trajectories are today applied to model the plague's history in Europe or to explain the recent spread of mad cow disease. His random hypothesis infuses all branches of scientific inquiry. You want to understand how the drug you swallowed will reach your cells? The answer can be found, at least in part, in Einstein's 1905 paper. You want to follow the spread of ideas or innovations? The solution is based on diffusion theory. In fact, randomness and diffusion impact everything from the design of nanomaterials to the marketing of new drugs.

Contemporary humans, however, pose a problem not faced by atoms, drugs, and our plague-carrying medieval ancestors: Today we can travel fast and far, dragging our diseases with us. Indeed, we hop into our car, and a few minutes later we are miles away. We can also board a plane, and when, hours later, we land, we have transported the virus with us to some other continent. Thus, if we want to predict the spread of infectious diseases we first need to answer a simple question: Can Einstein's theory still capture our mobility?

One could argue that modern transportation merely speeds us up, turning us into atoms on steroids that move faster and jump farther. The nature of our trajectory, however, remains unaltered, and we remain as unpredictable as the atoms. Hasan's trajectory—from the United States, to Amsterdam, to Lisbon, to Paris, and so on—surely supports this conclusion, underscoring the apparent arbitrariness of human motion.

We cannot exclude the possibility, however, that human trajectories are fundamentally different from those traced by atoms and molecules. And interpreting how individuals move and travel is surely more useful than anticipating the trajectory of an atom or a pollen spore. Not only would it help us halt the spread of the next deadly disease, but it might help us design better, more sustainable cities. Perhaps we could use it to develop computers that, foreseeing our whereabouts, could provide us with all the information we need, not now but where we will soon be.

Yet such advances require that we do for humans what Jean Perrin did in 1908 for molecules: design a way to track their motion. The problem is, nobody knows where we are and where we go; because, with the notable exception of Hasan, we do not spend our lives under a microscope. But Dennis Derryberry, the jack-of-all-trades cabinetmaker from Vermont, was quick to realize that, in WheresGeorge.com, we *do* have a way to trace human movement. Indeed, the main reason the money moves is because it travels with us. Thus Georgers like Gary, who meticulously record the circuitous passage of single dollar bills, are the twenty-first century's Jean Perrins, trackers of human mobility.

Soon after their chat under the Vermont stars, Dirk Brockmann packed his bags and returned home to Göttingen, where he introduced Wheres George.com to his colleague Lars Hufnagel and his boss Theo Geisel. Lars, a *Sauerländer,* known for speaking only when it's absolutely necessary, had little to say beyond "Hmm, very interesting," and proceeded to download each bill's trajectory from WheresGeorge.com. A few days later he showed Dirk and Theo the first analysis of the distances covered by the bills. He found that 57 percent of the bills stamped in New York were within six miles of their release point two weeks later. Similarly, after the same lapse, 74 percent of those released in Jacksonville, Florida, were still in their original neighborhood—out of a wallet, into a cashier's drawer, and back into somebody else's wallet, only to be spent again two streets away. This was hardly surprising and more or less consistent with the expectation that the bills travel randomly. And given the fact that banknotes travel only where we carry them,* the bill's random trajectory suggests that our paths are also unpredictable.

A few bills did not fit this random pattern, however. Two weeks after their initial circulation, about 7 percent of the New York bills and almost

---

* Bills certainly move about outside of our billfolds—occasionally they go from merchants to banks and are then transferred on to other banks or merchants. But such travel is negligible compared to the distances they travel in our pockets.

3 percent of the Jacksonville bills were found at least five hundred miles away. These traveled not only much faster than the rest but followed a pattern quite different from the one diffusion predicted: Their trajectory was dominated by a few extremely long jumps. Einstein's theory could easily account for the many small steps the bills took. But it failed to explain these rare but large superdiffusive jumps.

If you think about it, these large jumps are by no means surprising. Say you stuff your wallet with cash at an ATM in Queens, New York, before heading to Kennedy Airport for your flight to Seattle. Upon arrival on the West Coast you would be glad to thin your wallet by handing some of the cash over to the cabdriver. With that you have moved the bills twenty-eight hundred miles, restarting their random journey in Seattle's local economy. But Dirk and his colleagues in Göttingen were intrigued that these big jumps followed a pattern different from that predicted by Einstein's theory. A pattern that allowed them to make several fairly precise predictions.

A droplet of red dye falling into a glass of water leaves a mark where it lands. Within an hour or so the dye fully mixes with the surrounding water, at which point its time and place of entry are indiscernible. The dye mixes thanks to the water molecules' random kicking, and Einstein's 1905 theory tells us precisely how long it will take until the entry point vanished amid the newly tinted water. In the same fashion, Dirk's superdiffusive law allowed him to predict how long before we lose track of the origin point of a pile of dollar bills spent in Queens. This is a critically important discovery. Especially if you have a suitcase heavy with counterfeit money and don't want the FBI at your doorstep.

Dirk predicted that in sixty-eight days the coast would be clear. That is, in about two months your counterfeit bills would be all over the United States, leaving the FBI no way to trace their origin.

There was only one problem with this prediction: It was blatantly wrong. Indeed, Dirk's own measurements showed that most bills released in New York one hundred days later were still in the city's vicinity. Despite their superdiffusive trajectory, an invisible hand slowed the bills down, forcing them to defy the laws of superdiffusion and stay put in their original neighborhood.

Dirk Brockmann was not prepared for the attention that followed the release in *Nature* of his paper on the motion of a dollar bill.

"Banknote tracking helps model spread of disease," proclaimed the *New Scientist*'s cover on January 26, 2006, and the British *Guardian* offered its own headline: "Money talks: Tracking dollar bills helps explain how diseases spread." The story was covered by most news outlets, from CNN to the Chinese *People's Daily*.

The Georgers, who had never heard of Dirk, were even less prepared. Yet even those too busy stamping bills to check the news noticed something was up when WheresGeorge.com slowed to a standstill and then crashed altogether. The server was unable to cope with the flood of new users that followed the media exposure.

One might think that the Georgers would be upset by such a major disruption. On the contrary, a general sense of jubilation followed the crash. Jen, from St. Louis County, who had stamped 13,686 bills herself, crowed, "this is for all those naysayers who we have to defend ourselves on why we love doing this," adding, "No, it's not defacing currency, and *now* we're scientifically helping research!! It just doesn't get any better than that!" Tony, with his 3,560 bills in circulation, proclaimed, "I'm proud to be a Georger today!" a feeling echoed by Andrew from New York: "For years I have been laughed at for this hobby, and now I can point to this story as a practical use for what we do."

Another Georger, going by the screen name SueBee, had a more practical question on her mind: "I'm wondering if the next time a cashier questions me I can tell them it is part of a scientific study to track how disease might spread based on analyzing how/where a dollar bill might travel." This drew a quick reply from Mike from Pittsburgh: "I wouldn't recommend it. Unless you spend a lot of time explaining very carefully, they are apt to think the bills have been infected with disease!" This plagued Dirk as well, who found himself having to patiently explain to the press, over and over, that, no, they had *not* found that AIDS spreads on dollar bills.

֍

Einstein wanted to know how atoms move before anyone had even seen one. He had to wait until Jean Perrin found a way to record the herky-jerky motion of tiny particles punted around by the still-invisible atoms. Though Perrin could not see the atoms themselves, what he *did* see was sufficient to support their existence. In the same fashion Dirk Brockmann wanted to know how humans travel. He had no means to track each of us separately, but in finding a way to track our money he discovered a way to track our mobility.

Einstein's and Perrin's Nobel-winning work offered the definitive proof of the existence of atoms. Dirk's work, however, offered us our first real glimmer of hope that one day we might infer the laws of human mobility. In the spirit of good science, it also drew attention to an important mystery: Why is the money not spreading as fast as the mathematics of superdiffusive motion predicts? Is Einstein's approach wrong, or are we wrong to apply his theory, developed for atoms and molecules, to human activity? After all, humans are not atoms, reducible to simple equations; there is always an element of unpredictability to anything we do.

During the century that has passed since Einstein and Perrin published their work, the experimental tools have significantly improved and no longer rely on pollen to prove the existence of atoms—a new range of powerful microscopes allows us to see them directly. In the same fashion, with the exploding prevalence of mobile phones, GPS, and other handheld devices, there is an abundance of new tools capable of tracking human mobility. Thanks to these gizmos, today just about everything we do is under numerous "microscopes" of some sort. The data collected by the various devices we all use is not idle, however: Corporations use it to boost productivity and to track everything from shipments to deliveries. The government uses it to catch terrorists. Countless start-up companies build on it to exploit our whereabouts and behavior, hoping to become the next Google. We live in a data-rich world.

Putting these data to lucrative use propels the development of further technologies that aim to discover even more about each of us. As we will

see, some of these technologies reveal that the discrepancy Dirk uncov-
ered communicates fundamental properties of human behavior. And
this discrepancy will force us in the coming chapters to reexamine many
things we hold sacred about the nature of time and space and the way we
experience them.

But first let us leave behind the humble dollar bill and in an Einsteinian spirit let us jump across both time and space, all the way back to 1514,
to a city once called Nándorfehérvár, which is today better known by its
Serbian name, Belgrade. There we will meet the man whose haphazard
trajectory will come to dominate the rest of our journey.

4

Duel in Belgrade

Belgrade, February 28, 1514, eleven months after the papal election

hy fight one another needlessly?"* bellowed the Ottoman captain as he halted his cavalry, which moments before had been charging at full speed into a group of Hungarian knights bracing for a fight. As his men reined in their agitated mounts, they finally gave the Hungarians a good look at the Turk. He was a short man but broad, his expansive chest shown to advantage by the embellished plated mail. A domelike helmet partially obscured his face before continuing down into a gorget that fanned over his shoulders. Strong, fast, and poised, he was Ali of Epirus, the dreaded captain of Szendrő. Not one of the Hungarians failed to recognize the warrior—each man standing knew someone slain by him.

"Your land is barren, stripped by our fighting; there is nothing more to plunder," continued Ali. As far as the eye could see, uncultivated fields, broken-down fences, and dilapidated cowsheds bore witness to his claim.

---

* Thus spoke Ali of Epirus, according to the chronicle of Ludovici Tuberonis, *Commentariorum de temporibus suis libri*, 1603.

Only the magnificent white-stone walls surrounding Belgrade stood whole and strong behind them in the bright February sun. The fortress, the crown jewel of Hungary's southern defenses, was secured on the south and west by thick, imposing walls and on the north by the wide, wending Danube. The inner castle and palace were further protected by the upper town with the main military camps, the lower town with the urban center and the cathedral, and the port at the Danube, ingeniously separated from one another by trenches, gates, and towering walls.

Beyond the restless Turkish cavalry, about twenty thousand steps from Belgrade, rose the fortress of Szendrő, whose twenty-one bastions were modeled after the mighty fort in Constantinople. Szendrő had been erected to protect from the invading Turks the strategic delta at the junction of the Morava River with the Danube. In 1459, however, the fortress had fallen to the Ottomans, and now it was under the skillful command of Ali of Epirus, whose cavalry, chafing for a fight, was impatiently staring down the Hungarian knights.

"This desolation offers meager spoil," Ali spoke again, his voice mixing with the clangor of the cathedral's bells. Not one of the Hungarians was ready to cede any ground, desolated though it was by the bloody standoff between the two empires. This was no ordinary land they were defending; the fortunes of Christian Europe had turned on this very spot almost sixty years earlier. In 1456 Sultan Mehmed II had arrived at the walls with a massive host of two hundred ships and seventy thousand soldiers. Belgrade's paltry guard, numbering a mere seven thousand, counted only the strength of the formidable fortress in their favor. They had also pinned their hopes on a rescue mission, led by János Hunyadi, the voivode of Transylvania and governor of Hungary. Yet the threat posed by Hunyadi's growing power was more immediate to the nobles than were their fears of the Ottomans sweeping across Europe. And so the promised army was nowhere to be found. Despite Pope Callixtus III's call to arms, fearing the end of Christianity if Belgrade fell, the continent remained indifferent. The pope, in a distraught attempt to remind everybody of the imminent danger, ordered the bells to ring at noon to solicit prayer for the desperate defenders.

Hunyadi's only ally was an old Franciscan friar, Giovanni da Capistrano, who at the age of seventy preached the Crusade so effectively that thousands of peasants, armed only with slings and scythes, joined Hunyadi's camp. So, thanks to the rhetoric of an aging cleric, the army quickly swelled to almost thirty thousand. Comprised chiefly of unskilled peasants, they were not a challenge to the sultan's expert forces of seventy thousand. Yet Hunyadi, a gifted tactician, overran the Ottoman camp in a surprise attack, wounding Mehmed II and forcing the remainder of his vast forces into retreat.

As the news of the incredible victory reverberated across the continent, the pope ordered that the church bells continue their noonday peal, celebrating that Christianity was saved.

Today when church bells sound at midday, few realize it commemorates the defeat of the Ottomans. But a mere fifty-odd years after Hunyadi's victory, the meaning of the toll was not lost on the Hungarian troops, as they faced this fresh assault.

"Let us look instead to military glory," Ali called to the enemy. The bells rang on.

In truth, the Hungarians—fully aware that the small Ottoman cavalry, without artillery, could do little harm to the impressive walls hugging Belgrade—had no need to leave the fortress. Perhaps a mixture of pride and a sense of opportunity beckoned them from safety to confront the enemy. In any case, the decision to face the Turks rested with the unit's captain, a tall man with high cheekbones and a long horseshoe mustache, whose downward-pointing ends extended well beyond his bushy beard. Mounted on a strong horse, the captain wore a faded hauberk that covered most of his body. His helmet was an old steel pot lined with worn leather, and the escutcheon on his shield, an inverted crown supporting a helmet-covered head, was battered and faded.

Slightly behind him, yet within hearing of a whisper, his younger brother bestrode his own mount. The inseparable twosome had been christened the Székelys by their partners-at-arms, in reference to the brothers' eastern Transylvanian tribe. As a contemporary chronicler put it, the Szeklers "did not pay taxes to anybody, neither to the king, nor to anybody else," a privilege, in those days, extended only to the noble class. In exchange, they vowed to take up arms for their king or voivode each time the bloody sword—their traditional call to arms—was carried through their villages.

In fact, centuries of battles had turned the Szeklers into a military society, and even the humblest soul in a Szekler village was ready to gather his helmet and dagger or scythe if beckoned.

Given their roots, the skills at arms of the two Szeklers surprised few at the Ends, as the dangerous no-man's-land between the Hungarian and the Ottoman empires was called. It was less clear, however, why they had joined the mercenaries instead of living as noblemen within their traditional borders. As the two never attended a fight so as to have a good story to tell, they kept most of their fellow men-at-arms wondering about their presence at the Ends.

Despite the brothers' uncanny physical resemblance, their personalities were like night and day. György, the elder, was strong-willed and quick-tempered, making him a popular leader in this fist-and-sword-dominated land. Gergely, by contrast, was calm and thoughtful, considering carefully before ever raising his voice in decision. Though skilled with a sword, he had countless times in the past, by his wise council rather than the strength of his arm, saved György from his rashness. Many suspected that the brothers' presence at Belgrade might have to do with past incidents when György had ignored his younger brother's counsel. Yet no one knew for sure, as no questions were ever asked over here.

"If there is any man among you Hungarians willing to prove his worth and trusting in his own courage not to fail," roared Ali across the divide, "step up and fight me, one-on-one." The challenge sent a ripple through the Hungarian knights. As a host, buttressed by the impressive guard along the walls, they had little to fear. Facing Ali alone . . . well, that appealed to no one.

The silence following Ali's challenge lengthened, second by second, punctuated by the chime of the church bells. *Bong, bong, bong.* The reminders of glories past now mocked and shamed the silent knights, who studiously avoided meeting one another's gazes.

After scores of years and even more fights together, Gergely could almost feel his brother's blood churn. His muscles were tense and flexed, his body readying to spring into a fray. It would be suicide for György to meet the Ottoman's challenge, Gergely knew, but he was at a loss as to how he could hold his brother back. Outright admonition—"Don't do it; don't go"—would have exactly the opposite effect. But what to do? Finally, Gergely clenched his jaw and hissed through his teeth in a low voice, inaudible to any but his brother, "György, if one of us is going to die, it won't be me."

The captain did not flinch, but small wrinkles at the corners of his eyes evidenced the smile fighting to melt his stony face. There was a slight shift in his tensed posture; his muscles seemed to relax a fraction under the armor. György then nodded in acknowledgment. Relief. Gergely had defused the situation.

But then, suddenly, the silence was sliced by the scream of György's horse as his rider's spurs dug into his flanks. Gergely's reverse psychology, his teasing challenge to survive, had obviously failed. As the animal reared, György leaned forward, his head disappearing behind the horse's neck while he calmly kept the reins slack. The animal shot forward, every taut stride carrying him closer to the Ottoman.

Just as abruptly as he had spurred his mount, György drew him up, fixing the Ottoman in his stare. He then pulled his long sword from its scabbard and was acknowledged by Ali, who raised his curved, single-edged saber and positioned himself to meet the knight.

As if compelled by an invisible hand, the two warriors abruptly sped toward each other, and the sharp points of their swords whistled through the air. In seconds the distance between them closed, and they met with a terrific crack. Ali's saber struck György's worn shield with a blow that would have shattered any other arm. Simultaneously, the knight's sword swung and missed its target, sliding along Ali's mail.

The long sword's impact along his mail and the momentum from his own blow delivered to the Hungarian's shield knocked Ali off balance. Though he remained unharmed, Ali was unable to regain his balance. The weight of his iron armor pulled him to the dusty, trampled ground.

A cheer erupted from the Hungarian knights as they watched Ali fall and lose his saber from the impact. But as the Ottoman, with a speed that denied gravity, rose to his feet and reached for his saber, the cheers were swallowed just as quickly as they had begun.

While Ali was scrambling to regain his saber, György Székely had turned his horse and now urged him into a long, stretching leap, aiming to trample his foe. György dropped the reins and reinforced his grip on the hilt of his long sword, now holding it with both hands. He then leaned to the right, lowering his upper body, while his legs tensed and burned, clamping onto the horse's flanks.

For a brief moment, the tip of György's sword ran parallel to the ground. Then, just as Ali's fingers closed on his saber, György's sword found the vulnerable joint in the armor at Ali's armpit, first breaking open the mail and then cutting deep into the muscle. Before anybody could register the rapid sequence of events, Ali's right arm, severed from the shoulder, flew through the air, his fingers still tightly gripping the saber's hilt. The limb landed on the cold February ground with a dull thud, accompanied by the loud metallic sound of the saber blade clanking onto a rock.

5

The Future Is Not Yet Searchable

As television and computer screens turn into each other and multiply, giving rise to an ecosystem of more autonomous news and entertainment, as ideas, ordinary people, and events are pulled from oblivion, trained in the spotlight of public adoration, revulsion, or fascination, and then ultimately forgotten, Andy Warhol's celebrated statement—"In the future, everyone will be world famous for fifteen minutes"—is a closer description of reality today than it has ever been since he first uttered it in 1968. *Time* magazine acknowledged this trend in 2006 by naming you its Person of the Year. Indeed, it *is* you who, via YouTube, MySpace, Wikipedia, Facebook, and *American Idol,* have transformed a world previously captivated by headline geniuses like Einstein or Warhol into one mesmerized by ordinary people who grab the spotlight for their fifteen minutes. Fame has given way to *fameiness,* a term coined by a *Los Angeles Times* column to capture this fleeting me-centric celebrity.

The future is not yet searchable. We perch tensely on its edge, absorbing the news that represents the boundary between our past and our future. Given our voracious hunger for the next big thing, any event that grabs headlines now is fated to soon fade. A story may start the news cycle prominently posted on the front page, but as new developments stream in, it quickly and inescapably loses its newsworthiness.

This is, of course, part of the social contract between the public and the media. Much like the snowflake that melts on contact with a warm finger, news matters only when it touches you for the first time. News must be news, and to keep up we feel obliged to check regularly, making yesterday's paper instantly worthless. You must read fast, as this chapter may soon become irrelevant. . . .

But did Warhol have it right? Is it really fifteen minutes? Or is it actually as long as half a day? Perhaps as short as five minutes? How long will a piece of news remain news in today's information-hungry world? Can we truly measure the fleeting moment of fame? Perhaps these questions sound silly to you. But they are by no means inconsequential. And as we examine them through the lens of science and alchemy, they will reveal a mysterious oddity about human activity.

∼

The year after Gary Kanis began stamping dollar bills, I left the United States for Hungary to join the Collegium Budapest for a sabbatical. The institute is housed in an imposing sixteenth-century Italian Baroque building in the heart of the Buda Castle that once was Buda's city hall. The building lost its tenant when Buda and Pest, the cities on the right and left banks of the Danube, merged in 1873, giving birth to the city of Budapest. In 1992 the building became home to a newly founded Institute for Advanced Studies, an elite research institution that continues to attract about a dozen visiting scholars from all over the world each year.

One of the fellows, a prominent Romanian architect who designed countless Orthodox churches, explained to me that in the past churches were always built on sacred sites. Mountaintops or hills were preferable, as they brought the congregation closer to the divine. Similarly auspicious were major intersections, crossroads, and city centers, where the community's paths naturally crossed. Based on these criteria, perched at the highest point of the hill that loomed over Hungary since 1265, the building that is today the Collegium should have by all rights been a church. With its thick walls and deep windows lending a monastic air to its many spacious offices, the Collegium is beyond inspiring. It feels almost sinful to sit within its walls and not at least attempt to dream on a larger scale.

My sabbatical coincided with the release of the Hungarian version of my previous book, *Linked*. The translation had been sponsored by the Internet spin-off of Hungary's formerly state-owned phone company. Its CEO, György Simó, is a sociologist who, after graduating from college, helped run Hungary's first, and initially underground, community radio station. In 1997 he joined the content group of the phone monopoly and created Origo.hu, which soon became the largest Internet portal in the country. So the spring of 2003 found me dining with Simó and two of his friends, chatting about *Linked* and interconnectedness.

By that time I had become passionate about the next great questions for network science: How are networks used? When is a Web site visited on the World Wide Web? When and how do people interact? Unfortunately, the necessary data was chiefly collected by large companies who jealously guard it as a trade secret. Indeed, Internet providers and search engines from Google to Yahoo! record at an incredible rate what we do, and when we do it. But the rapid monetarization of this data has sidelined scientists interested in exploring questions of basic human behavior. Today Google is the poster boy of this trend; its famous "Don't be evil" philosophy covers a huge intellectual black hole. The company uses its billions of dollars to sweep up the best engineers and scientists in their fields, who are then locked up in its Santa Clara Googleplex, where they are shielded by strict nondisclosure agreements so that they can rarely publish their findings.

That spring night in 2003, Simó, the former academic and samizdat writer, listened and offered help. The next day he introduced me to András Lukács, a mathematician who was archiving the Web-browsing patterns of the people who visited Origo.hu. He did not know who these users were, but every time a person arrived at the site he knew which articles she read there and how many minutes she kept at it before giving up and clicking elsewhere. The extraordinary number of clicks Origo.hu attracted each day made it virtually impossible to follow any one user in real time. But months later the database offered a glimpse into the millions of clicks poured on the site. Lukács and his collaborators removed identifying personal information from a month's worth of records, representing more than 6.5 million hits, which in turn was about 40 percent of Hungary's Internet traffic at that time. With that they offered us access to the browsing habits of nearly the entire nation.

Armed with Lukács's browsing records, two researchers in my group at the University of Notre Dame—graduate student Zoltán Dezső and postdoctoral associate Eivind Almaas—set out to answer this simple question: How long is any particular news item accessed at a given time at Origo.hu? In other words, how long is each story's fifteen minutes of fame *really*? To answer the question, they first determined the number of people who clicked on a particular article each hour. Not surprisingly, about 28 percent of traffic occurred during the initial twenty-four hours after an article was published online. On the second day, there was a dramatic drop-off in readership, accounting for a mere 7 percent of the article's total hits.

This makes sense, of course: If a piece of news interests you, you tend read it as soon as you come across it while trawling your favorite sites. And by the third or fourth day everybody who had any interest in the piece had a chance to look at it, thus the visitors should have ceased arriving. The problem was that they did not. Instead most articles on the portal continued to be read many times over, many days after their initial publication.

This was somewhat mysterious. First of all, how did people even come across these articles, many of which had disappeared from the front page days prior? Second, why on earth would anybody be interested in old news?

The answer to the first question proved simple: Some articles found a second life in popular forums and via links from other Web sites. The second question proved more stubborn: Why *would* people regularly click on certain news items that were anywhere from a week to a month old?

ҿ

In [Saturn] is hid an immortal soul. Untie its fetters which do it forbid to sight for to appear then shal arise a vapour shining like pearl
orient. To Sa[-]turn mars with bonds of love is tied who is by him
devourd of mighty force whose spirit divides saturns body & from
both combined flow a wondrous bright water in which the sun doth
set & lose its light. Venus a most shining star is embrac't by [mars].
Their influences must be united for she is the only mean between
the Sun & our true argent vive to unite them inseparably.

The author of this 1670 fragment, Jehova Sanctus Unus (Jehovah the
Holy One), appears to reflect on the role that planets play in relationships
and love, a favorite subject of astrologists even today. Yet all is not as it
seems. The Saturn in question refers neither to the planet nor to the
Roman god but is actually a code word for *stibnite*, the key ingredient for
the metalloid element *antimony*. Neither is the *immortal soul* a spiritual
reference but a stand-in for antimony itself, a volatile component that
mysteriously evaporates once heated. Venus means copper, Mars iron,
and Saturn's devouring Mars refers to the chemical reduction of antimony from stibnite by means of iron. Finally, the phrase *wondrous bright
water in which the sun doth set* is not a poetic riff on glorious maritime
sunsets but the dissolution of gold in sophic mercury.

The excerpt is a riddle, a poetic reflection not on astrology but on its
cousin, alchemy. It encrypts the secret formula for the universal agent of
transmutation, *the philosopher's stone*. Throughout the Middle Ages alchemists believed that such a stone could rid any metal of its impurities,
turning it into the untarnishable and thus immortal gold. It could also
cure human illness, so it became the ultimate symbol of humanity's
magic power over nature.

While Jehova Sanctus Unus may not be a household name today, he
was in his own time an intellect nonpareil. He had cracked nature's code;
his reputation prompted Einstein to call him the "inventor of genius"
who "deserves our deep veneration." So who is Jehova the Holy One, pos-
sessor of the *prisca sapientia*, the secret wisdom passed down from Moses
through a line of illustrious successors, including Pythagoras and Plato?

In 1936 the venerable auction house Sotheby's put up for sale 329 lots of manuscripts, more than a third of which meticulously detailed attempts at creating a philosopher's stone. The manuscripts had been labeled NOT FIT TO BE PRINTED following the death of their author in 1727. The secrecy was well-advised, as the documents revealed the true identity of Jehova Sanctus Unus, the man whose lifelong obsession was to transmute ordinary metals into pure gold. He was none other than Sir Isaac Newton.

To the best of our knowledge, Newton never succeeded in turning antimony, lead, or any other metal, for that matter, into gold. Even so, this period of his life was singularly productive; between his experiments in alchemy he wrote the *Philosophiae Naturalis Principia Mathematica,* a work of profound mathematical insight, unraveling his discovery of the laws of gravitation and planetary motion. The aims of his life's work, the laws of transmutation, however, eluded him and were only deciphered two centuries later.

In 1901, as Frederick Soddy observed radioactive thorium spontaneously converting to radium, he cried jubilantly to Ernest Rutherford, his colleague, "Rutherford, this is transmutation!" Not particularly keen on his colleague's take of their discovery, Rutherford is said to have snapped, "For Christ's sake, Soddy, don't call it transmutation. They'll have our heads off as alchemists." Eventually, their discovery merited a Nobel Prize in chemistry and proved once and for all that one element can indeed change into another without human agency.

Though shortly before Soddy and Rutherford's discovery Pierre Curie and Henri Becquerel had observed radium turning into uranium, Rutherford won the Nobel for successfully quantifying his observations, deriving the laws of transmutation. His insight was simple: Radium atoms are all identical and indistinguishable; therefore, at any moment in time, each is equally likely to turn into thorium. That is, the moment that a radium atom, driven by some "incurable suicidal mania," turns into thorium is inherently random. This hypothesis allowed Rutherford to predict that the number of radium atoms will decrease in time following an exponential law, in excellent accord with his experimental data.

✌

So, back to Hungary and the Web portal Origo.hu. Imagine that each visitor who has not yet read an article is actually a radium atom that has not yet disintegrated. Driven by some incurable curiosity, they click on links to news items at times that to us, not knowing each visitor's schedule and motivations, appear random. Along the way they create a pattern quite similar to the random disintegration of radium atoms. Thus, starting from Rutherford's theory, we were able to derive a mathematical relationship that predicts the number of Origo.hu users who have not yet read a news item they were interested in. The calculations indicated that the number of potential readers should decay very fast, exponentially in fact, as predicted before by Rutherford's law of transmutation.

The beauty of the theory was its simplicity: It required only a single parameter—the average number of times a user clicks on the site each day—to predict the visitation history for each article. This parameter was easily obtained, as the data set indicated that a typical user clicked on the site about twenty-six times each day. With this in mind, Rutherford's model, adapted to the Web, made a rather specific prediction: Thirty-six minutes after the publication of a news item, more than half of the visitors who will ever read it will have already clicked on it. That is, the half-life of a news item should be about thirty-six minutes.

Thirty-six minutes isn't a far cry from fifteen minutes. Indeed, it's a pretty good approximation. So, a combination of meticulous measurements and theorizing actually confirmed Andy Warhol's insight about fame. Never mind that Warhol talked about a person and we focus on a news item here—he would have drawn little distinction between the pop starlet herself and the gossip she generated. Yet how did Warhol intuit today's behavioral patterns with such a degree of accuracy in 1968? Was it by mere coincidence, or could there be some deeper truth behind his prescience?

In the end, our prediction did not hold. Indeed, we soon realized that we do not have to resort to theories and models but can directly measure the half-life of each article posted to the Web site. The results were dispiriting: Not only did the measurements fail to corroborate Warhol's fifteen-minute theory, but they disagreed with our thirty-six-minute prediction as well. Instead, it took about twenty-one hundred minutes, or thirty-six hours, for half of the users to click on a typical article. Thus, our prediction and Warhol's insight were both orders of magnitudes away from reality.

Who cares, you might ask. And maybe you're right—not being able to pinpoint the number of hours we devote to Paris Hilton's string of misadventures should perhaps come as a blessed relief. Why would any self-respecting scientist worth his salt care about such a silly thing anyway?

Granted, were Paris Hilton or Andy Warhol our only concern, the discrepancies would not have troubled me. And yet I *was* concerned, and not because I really needed to know the half-life of any Origo.hu article—direct measurement had already established it at thirty-six hours. No, what disturbed me was that the disagreement between the predictions and the measurements meant that our understanding of the Web-visitation process was inadequate. Grossly inadequate, in fact. And in turn, it also meant that our comprehension of human behavior must be seriously limited.

≈

Rutherford is famous for his arrogant oft-quoted assertion that "in science there is only physics; all the rest is stamp collecting." To be sure, the search for basic principles and laws, which is the ultimate quest of physics, is the foundation of many spectacular advances in our time, from the transistor to space travel. Yet, it is an approach that does not fare well in many branches of science. Biologists confronted with the mindboggling complexity of a cell, brain scientists humbled by the miracle of our neural circuitry, and social scientists and economists at a loss before the labyrinth of social and economic processes, from economic bubbles to crises, have often argued that fundamental laws may not exist in their fields. As such, many judge the physicist's insistent search for universal laws to be misguided at best and, at worst, bound for failure.

These skeptics would be consoled by the failure of each of the two research projects we have encountered so far in this book—Dirk Brockmann's discovery of the laws governing human travel and our quest to characterize the laws governing Web visitation. Indeed, they confirmed the physical method's limitations when applied to human behavior. While Rutherford's exponential law captures transmutation with amazing accuracy, it fails to account for the decay of Web visitation. Similarly, Einstein's diffusion theory captures the random motion of atoms yet fails to correctly predict the circulation of banknotes.

Equally puzzling are the parallels between the two failures. In both cases our predictions were too fast. In reality the dollar bills spread much more slowly than predicted by Einstein's diffusion theory, and Web site visitation was much slower than predicted by Rutherford's model. That is, both projects hint at some invisible hand that slows human-driven processes. These discrepancies between prediction and calculation are frustrating at least. But more critically, they beg the question, can we ever hope to describe human behavior as accurately as we are able to describe the material world?

❧

Today an Illinois-based company will convert the cremated ashes of your loved one into a 1.5-carat diamond for a mere $24,999. While I am uncertain whether or not I really want to see my grandma in a diamond ring, one has to admire the marketing audacity fronting this technological feat. Strictly speaking, this is not transmutation, as ashes and diamonds are both composed of carbon. Not bad, though. In 1980, Glenn Seaborg, following in Rutherford's footsteps, successfully made gold out of the chemical element bismuth. The feat would have certainly made Newton proud. But Seaborg found the procedure so energy-consuming that it was hardly viable economically.

Seaborg's success did demonstrate, however, that quantitative methods can be applied to areas that science has long considered dubious or even fantasy. Could it be that if we refine our scientific methods we can one day achieve similar metamorphoses with human behavior? Could we turn it into an accurate, predictive science? Could we stop the next pandemic by auguring the virus's path, telling you exactly what street to avoid tomorrow to avoid contagion? Might the evening news stop reporting events past and instead begin operating more like the weather guy, informing us what we should expect in human affairs in the days to come?

As fundamental as these questions are, they remain so far beyond the current grasp of academic enterprise that most scientists prefer to steer clear of them altogether. It is not even clear which fields of science could best address them, were they able. Physics? Biology? Economics? Computer science? Psychology? Any number of the social sciences? And so, predicting human behavior is presently left to business consultants and palm readers.

Bogus though their predictions may be, fortune-tellers and prophets don't suffer for business, having determined the fates of kingdoms, Crusades, and, more recently, business empires and entire economies, as we shall see in the following two chapters.

6

Bloody Prophecy

City of Buda, the royal palace, March 24, 1514,
three weeks after György Székely's duel with Ali of Epirus

f the pope's wishes are announced to the people, I must agree that many will gather when summoned," said István Telegdi thoughtfully.* He turned, as deference required, to Ulászló, king of Hungary, but he carefully projected his words toward the surrounding nobility. The dark, dank walls of the neglected throne room still echoed with the voice of Cardinal Bakócz, who had just conveyed the pope's message to the court.

Two years prior, King Ulászló and his confidants had gathered in that very room to bid farewell to the cardinal, who was setting out for Rome in hopes of being elected pope, successor to Julius II. Counsel, support, prayers, and especially money accompanied the first Hungarian pope-to-be. But today, almost a year had passed since the court had learned that the papacy had instead gone to Giovanni de' Medici, the young aristocrat from the powerful Florentine dynasty, now crowned Pope Leo X. The court was still

* After Georgius Sirimiensis, Epistola de Perditione regni Hungarorum (1545–1547).

perplexed: How could the conclave have passed over Bakócz—a master of international politics and intrigue—for a young, pampered courtier?

Lately the cardinal had become obsessed by how close he had come. To some he confided the sweetness of the moment when he realized that he had won over his rivals, only to realize that in his victory he had in fact lost the papacy. Unbeknownst to him, he had been fighting the long hand of the recently departed pope.

On his deathbed, Pope Julius II had shared his last wish with his confidants: Elect anyone *but* the Hungarian. So the real danger that Cardinal Bakócz might ascend to St. Peter's throne forced Cardinal Castellesi to ally himself with his longtime enemy, Giovanni de' Medici, and a compromise was born. In the conclave Bakócz had as many votes as he could count on, but he was short of the two-thirds majority he needed to become the pope. The young cardinals wanted one of their own—young, Italian, and guaranteed not to last, so that some among them would again have a chance at the bishopric of Rome.

Medici, carried in a sedan chair while convalescing from a recent surgery, fit the bill. A short reign was likely. Thus the morning following the first scrutiny, he had been elected pope. It was only after the white smoke wafted into the Roman skies that the cardinals noticed a minor procedural oversight: Despite having received his red, broad-brimmed galero three decades before, Medici had never actually been consecrated as a priest. Easily remedied: The next day he was ordained, and the day after saw his consecration as bishop. And finally, on March 19, 1513, Giovanni de' Medici was enthroned as Pope Leo X.

"God has given us the papacy; let us enjoy it," proclaimed Leo X. And he certainly did so.

But now the new pope's wishes have created quite a stir in the Hungarian court. So Telegdi continued, talking about those who might respond to the pope's call: "But who and what kind? The wretched, guilty of capital offences, those cast out of their homes for crimes?" Turning now to the cardinal, he added. "Outlaws, those ridden with debt, pimps, the rabble, marked with disgrace and exile, those with hope for nothing else in life but to sink farther into depravity and heinous crime?"

Telegdi, the king's trusted diplomat, was unaware that the cardinal's return to Buda had been neither his wish nor his idea. After the fiasco at the conclave, Bakócz had resigned himself to living out the remainder of his life in Rome as the Vatican's most influential cardinal. Yet a few months into his self-imposed exile he was given a critical mission, perhaps the most important of the century: Conquer Constantinople.

Leo X wanted the jewel of Eastern Christianity back from the Ottomans, who had taken it five decades earlier. Hence a new Crusade. And who better to lead the charge but the cardinal from the country with centuries of experience barring the Ottomans from Europe?

So now the cardinal was back in Buda, this time having arrived as an apostolic delegate, carrying a papal bull endowing him with authority over *"Hungariae, Bohemiae, Poloniae, Datiae, Norvegiae & Sveciae regna, nec non Prussiam, Russiam, Livoniam, Lithvaniam, Valachiam, Slesiam, Moraviam, Lusatiam, Transilvaniam, Sclavoniam, Dalmatiam, Croatiam & Moscoviam"*—that is, most of Christian Europe. It allowed him to raise armies and collect taxes for the campaign, to offer full indulgences and bless in heaven all who fell defending Christianity. In normal times only the pope held such privileges. But as the king of Hungary and his confidants among the court had just learned, the present times were anything but normal.

Despite the unprecedented powers bestowed on the cardinal, an important consideration remained unresolved: A vast army was required to conquer Constantinople. But where would it come from? A quick survey of the shabby throne room attested to the king's inability to lend significant coin to the cause. Now left to the ravages of time, the once-magnificent Buda Castle had been designed and built by architects brought all the way from Italy by King Matthias, son of the great János Hunyadi. But the buildings that King Matthias had started but did not live long enough to see completed were now mere ruins. Even inside the palace, the seat of the country's power, signs of deterioration were everywhere. The courtyard's magnificently sculpted water fountains were dry, and weeds grew out of cracks in their bases; broken windows had never been replaced, and the rendering fell from the walls, gathering in small, dusty heaps. Reminders of past glories filled the castle, dulled, unkempt, and abandoned.

And money for a new army? One thing was certain—that would not be forthcoming from Ulászló's coffers. The broad powers offered in the papal bull were largely symbolic, impossible to enforce in the surrounding kingdoms. Requests for support had been sent to Venice and other countries threatened by the Ottoman Empire, but it would be weeks before Bakócz could reasonably hope for a response. This begged the question why the cardinal, fully aware of the unpromising circumstances, would ever agree to a Crusade without the necessary army and resources. Because he had a plan.

Still vivid was the memory of János Hunyadi, who had held back the Ottomans' advance in 1456, saving all of Christendom from their onslaught. He and his band of ragtag peasants, outnumbered and outtrained, defeated the powerful Mehmed II at Belgrade. Cardinal Bakócz was merely requesting a repeat performance. In the name of the pope and holy Crusade, call the peasants to arms. Teach them to fight. Send them to wrest Constantinople back.

Telegdi, a rich landowner himself, was not shy about voicing his concerns over the plan. "I am sure that many peasants will be quick to assemble when summoned," he said. "But might they join the host merely to avoid the backbreaking effort to work the land or to avenge their many injustices suffered or to escape punishment and torture for crimes committed?"

Throughout the fourteenth and fifteenth centuries, Hungary had enjoyed a population explosion, doubling the number of its inhabitants to approximately three and half million. Most were serfs, peasants tied to the land. Since the nobility were exempt from taxes in exchange for military service, the serfs had increasingly borne the crippling levies. Many tried to flee to cities, where the taxes were more bearable, which only prompted the nobility to restrict their movements. Thousands of peasants worked Telegdi's vast holdings, and he was acutely aware of the hardship they endured. Though the king's counselor spoke convincingly, the cardinal remained convinced that the best tonic for the peasants' grief was to march them off to distant battlefields, far from their present suffering. Telegdi, however, was not the only one in the court who needed convincing.

Cardinal Bakócz's unanticipated return from Rome had upset the careful balance of power in King Ulászló's court. In the two years since the cardinal had left Hungary, his power had been assumed by the young prince János Szapolyai. After King Matthias had died, the three-year-old János was lifted from his crib by his father, then governor of Vienna, who had confidently proclaimed, "You will be king one day." Now at only twenty-four, János Szapolyai was the most powerful landowner in the country, and as the voivode of Transylvania, he ruled over a majority of the territory loyal to the Hungarian crown.

Previously, the cardinal had proved a shrewd interpreter of both sacred and profane laws, managing to keep the young voivode at an arm's length from the throne. But two years is an eternity in political maneuvering, and those loyal to Szapolyai had long since assumed the most powerful positions in the Buda castle. The king wasn't much engaged in the court's wrangling, consumed as he was in mourning the death of his beloved young wife and now both fathering and mothering their two young children, the heirs to the throne. Indifferent to the country's affairs, anxious and indecisive, he had been nicknamed "Dobre," from the Polish word for *good*, the one he responded with every time he was asked to sign a document.

With a weak king, the decisions once left to the cardinal were now Szapolyai's. And it was no secret that the voivode opposed this new Crusade. He had no love for the Ottomans—in fact, while Bakócz, Telegdi, and the courtiers debated the pope's decree, Szapolyai was off at the Ends doing battle with them. No, he opposed the Crusade because it would reinstate the cardinal's influence; an outcome neither he nor any of his allies wanted to entertain.

In Szapolyai's absence those loyal to him could be counted on to check the cardinal's power grab. Even so, Szapolyai's resistance was running smoothly, since Telegdi, very much his own man, seemed also troubled by the cardinal's proposal.

"And if, then, the nobility complain that the fields have been left to the weeds, the socage and all to which the peasants normally attend now left neglected; if the nobility reproach us for the damage this Crusade causes to them to suffer; and if they begin to imprison their serfs to prevent their even setting off for the military camps, or demand the return of the departed as fugitives, or, as so often happens in such cases, imprison the family and relatives of those gone to battle (and we should in-deed remember our own crimes, bitter witness to our greed); if this comes to pass, this mob, this shoddy crowd, now in arms, will you be able to keep them on mission?" thundered Telegdi. He then concluded his argument with an apocalyptic vision: "Will the armed peasantry not turn on and attack the nobility, freeing their own from the wrong they now suffer? The sword, given to them to destroy the enemy, will they not turn it against us, only to be desecrated with the blood of our wives, brothers, sisters, children!"

Telegdi rested. The cardinal discreetly surveyed the room to weigh the damage this dark oration had inflicted. Telegdi was too rich to be bought and was respected for his genuine concern over the fate of the kingdom. It was this impartiality that lent his arguments even greater weight. His willingness to speak his mind gave voice to the questions and concerns troubling everyone present.

Would he prevail, wondered the cardinal. Could he end the Crusade before it even began? At seventy-three, his recent prospects dashed, Bakócz had few options left, and he would train all of his skill on one: making the Crusade a success, at any price, and then returning to Rome the conquering hero, renewing his chances to ascend St. Peter's throne.

7

Prediction or Prophecy

The one that hits you, you'll never hear. So each scream of a shell was a relief; this one was not for them. The Prof, however, seemed oblivious to the explosions that periodically shook the shelter and calmly carried on with his attempts to predict that afternoon's weather.

His progress was slow, often interrupted by trips to pick up wounded from the battlefield. Though he wasn't in combat, his missions were not without danger; a few days earlier, when they'd entered a forest to rescue an injured gun crew, a shell landed right on top of their ambulance. Miraculously, no one had been hurt, but the close call had shaken everyone.

But yesterday's terror was forgotten as the Prof examined the weather reports for 7:00 A.M. and worked furiously to predict the weather six hours later, at 1:00 P.M. One detail, however, made his efforts possibly ridiculous: The weather he had been attempting to forecast was for the peaceful afternoon of May 20, 1910. Yet, today's calendar most resolutely declared the year to be 1917. As World War I raged around him, his comrades on the French Front shook their heads and wondered, *Who cares?*

Lewis Fry Richardson was a tall, quiet, affable Quaker who loathed any kind of violence. He had joined the British Meteorological Service in 1913 and soon familiarized himself with the weather-prediction techniques favored by his colleagues, more of an art than a science. The forecasters each day would examine the current weather map and find a day on record in the past on which the weather conditions were most similar. "The forecast was based on the supposition that what the atmosphere did then, it will do again now," explained Richardson.

Educated as a physicist, Richardson was not in his job long before he realized that there had to be a better way to do this. By 1913 the equations describing the motion of air as a fluid were already known. Therefore, if one knew the *present* weather conditions, in principle he could predict all subsequent changes in the atmosphere. Which meant that he could forecast tomorrow's weather, using only physics and math.

Richardson had outlined his new methodology in a book, completing the first draft the year prior. But before he sent it off to press, he wanted proof that his methodology worked. So, starting from a detailed survey of weather around the globe at 7:00 A.M. on May 20, 1910, he set out to methodically forecast the weather six hours later, at 1:00 P.M. But then World War I broke out. Though his pacifist religious beliefs exempted him from military service, he still enrolled with the Friends' Ambulance Unit, comprised of fellow Quakers, who transported the wounded from the battlefield. So between intervals piloting ambulances around the Western Front, Richardson doggedly continued his weather prediction.

It is often said that it took Richardson about six weeks to predict the weather six hours later. The truth is that it would have taken six weeks if he had worked at it continuously, twenty-four hours a day. In the end, given the many interruptions he suffered, the calculations absorbed the greater part of his two-year stay on the French front. Despite the exorbitant amounts of time and effort, the results proved rather disappointing. Where he had predicted a pressure change of more than 145 millibars, the observed variation had only amounted to 1 millibar. Given that the largest pressure change ever recorded in Britain was less than 130 millibars in a similar period, his prediction was in massive error. A bit like predicting snow for a typical balmy day in August.

Yet it wasn't only his accuracy that raised eyebrows; the resources required to compute his predictions were even more troubling. Long before the invention of the computer, Richardson envisioned a forecasting factory—a huge hall with a map of the world painted on the floor, filled with human "calculators," individuals skilled in math dedicated to solving equations pertinent to the weather in the region of the world where their desk sat.

Richardson estimated that to keep ahead of the changing weather, his forecast factory would have to employ sixty-four thousand individuals. This estimate proved to be as accurate as many of his other predictions, as he was off by 136,000, give or take. No wonder that no one attempted ever to put his forecasting factory into practice. After it had garnered a few polite reviews in the scientific literature, Richardson's dream, the first attempt to turn weather prediction into an exact science, was simply forgotten.

～

Though István Telegdi's run-on speech to the Hungarian court had been somewhat circuitous, its meaning was clear: If you assemble the peasants and give them weapons, they will turn on their oppressors. Their enemy is not the Ottoman but the aristocracy, the insatiable landowners. Pity Constantinople now, and you will fear for your life tomorrow.

Was Telegdi a clairvoyant, privy to visions of the future? Or was he merely only a tired, troubled old man, fearful of disruption and change?

Earlier we encountered two other forecasts of quite a different nature. Dirk Brockmann predicted that within sixty-eight days the bag of cash spent in New York would spread so widely that no one would be able to trace its origin; and we predicted that the lifetime of any news story is about thirty-six minutes. Neither was a prophecy but, just like Richardson's weather forecasts, a prediction based on mathematical models. But both failed when tested against reality. The money traveled much more slowly than diffusion theory had foreseen, and the lifetime of the news story was thirty-six hours, not thirty-six minutes. So if we consider Telegdi's prediction, Brockmann's estimate, Richardson's forecast, and the fruits of my own research, three are wrong, and one, Telegdi's prophecy . . . well, you'll see what happens.

We tend to be either believers or skeptics when it comes to predictions. The believers support a rich network of fortune-tellers, psychic readers, and business consultants and fuel the sales of the writings of the sixteenth-century prophet Nostradamus, whose riddles predict all major events in human civilization. The skeptics, however, argue that Nostradamus's ambiguity renders his visions void of any genuine predictive power and then point to frequent forecasting blunders by experts.

Take, for example, the forecast provided by the National Association of Realtors, whose economists reminded us that over the course of a year median home prices had "never declined since good record-keeping began in 1968," predicting a 6.1 percent increase for 2006. That year, however, home prices fell 3.5 percent, a sign of the burst to come in the U.S. housing bubble. You would think, embarrassed by their blunder, that the prognosticators would have adjusted their models. But guided by their own peculiar logic, they did not, as indicated by their optimistic December 9, 2007, press release: "Existing Home Sales to Trend Up in 2008!" Once again, the truth failed to cooperate, and December 23, 2008, found sales down by 11 percent, the worst housing slump since the Great Depression.

In light of these failures to forecast accurately, let's put the question again to the skeptics: Why couldn't Telegdi, a close observer of the social and political realities of his time, predict the outcome of the Crusades? After all, humans derive evolutionary benefit from accurately foreseeing events, devoting many brain circuits to it. When playing tennis, I move toward the tennis ball on the court as I instantaneously predict the necessary time, place, and velocity with which my racket must hit the ball in order to return it to my adversary's side. Similarly, seeing a rapidly approaching car, I can easily forecast that it will run me over if I time my walk across the street poorly.

There are good reasons to accept some predictions while rejecting others. We internalize patterns and thus accurately predict events that obey the laws of nature—like the flight of a tennis ball or the trajectory of a moving car. It is much harder, on the other hand, to foresee the outcome of battles involving tens of thousands of peasants, events driven by the conflicting interests of kings, cardinals, sultans, and voivodes. Dependent on human will, these events are difficult to foresee. But *difficult* is not the same as *impossible,* so as we pursue the science of human behavior, we must ask an important question: Can we *in principle* predict our future?

Our question, though critical, is hardly original, having been asked by none other than Karl Popper, one of twentieth-century science's most famous philosophers. He calls the expectation that social sciences will make historical predictions as the *historicist doctrine,* saying that it is "one of the oldest dreams of mankind—the dream of prophecy, the idea that we can know what the future has in store for us, and that we can profit from such knowledge by adjusting our policy to it." In his aptly titled 1959 essay "Prediction and Prophecy," Popper poses the following: "It is a fact that we can prophesy solar eclipses with a high degree of precision and for a long time ahead. Why should we not be able to predict revolutions?"

Popper leaves no ambiguity about his views, which set the agenda of social sciences for decades to come: If humans are involved, prediction is impossible, so don't even bother. His argument is simple but persuasive:

> Eclipse prophecies, and indeed prophecies based on the regularity of the seasons . . . are possible only because our solar system is a stationary and repetitive system; and this is so because of the accident that it is isolated from the influence of other mechanical systems by immense regions of empty space and is therefore relatively free of interferences from outside. Contrary to popular belief, the analysis of such repetitive systems is not typical of natural science. These repetitive systems are special cases where scientific prediction becomes particularly impressive—but that is all.

True, history is not repetitive, and neither are our motivations and desires: We always aim for better, for more, for something different. So, given Popper's authority, in 1959 the issue was closed before it was really opened for debate: When it comes to history or social sciences, prediction is not possible. Telegdi's auguring is farcical, and any attempt on our part to harness predictive power is bound to fail.

◆

Today we have a global-weather prediction system that over the past five years has been 95 percent accurate in its three-day forecasts. Surprisingly, perhaps, this impressively successful system is closely predicated on the methodology Richardson outlined in his book, making us wonder, why did his predictions fail so spectacularly?

The problem wasn't his method but his data. Today's weather models feed on detailed radar and satellite maps, surface and upper-air temperature updates from thousands of locations around the world, not the spotty collection of atmospheric conditions that was Richardson's starting point. Furthermore, fast computers have replaced the two hundred thousand or so human calculators he required, also circumventing certain mathematical instabilities that hampered some of his calculations. Does this suggest that our ability to predict the future is limited only by the quality of our data and the speed of our computers? Or is it, perhaps, that, regardless of how much information we compile or the speed of our processors, when it comes to human activity, we are bound to fail predicting it?

This is quite an Einsteinian dilemma. Indeed, the first paper Einstein mentions in his 1905 letter to Conrad Hebrich—the paper for which he will later receive his Nobel—is the foundation of quantum mechanics. Today quantum theory is without question the scientific theory with the most accurate predictive power ever invented by humans. For example, the magnetic dipole moment of the electron is correctly predicted by quantum theory to within about one part in 10^{10}, a precision unprecedented in the history of science. Furthermore, according to some estimates, currently as much as 30 percent of the U.S. gross domestic product is based on technologies made possible by the quantum revolution, from mobile phones to iPods. Despite this, paradoxically, quantum mechanics is inherently unable to predict future events with certainty. In most cases it provides only the *probability* that a given event will take place. True, for a massive object, like the hardcover edition of this book, the probability that it will vanish while you are reading it is practically zero. But when it comes to one of the electrons comprising this book, things are no longer quite so certain; it could very well vanish from here and reemerge on the other side of the world. Not that you would notice, but such an event is well within the realm of possibility in the quantum universe.

This probabilistic framework was so troubling for Einstein that later in his life he rejected quantum mechanics and, up until his death, continued to look for a theory that accessed a deeper reality, one not tethered to probabilities. He famously offered that "God does not play dice," a central thesis of his new intellectual crusade. He dreamed of a fully deterministic theory, like Newton's mechanics, unambiguous about future events.

Was this a quixotic quest of an aging colossus, similar to the cardinal's struggle to get his Crusades marching to Constantinople? Or could there be indeed fundamental limits to human predictability that we cannot overcome with better theory, more data, and faster computers?

Today we know that Einstein was wrong and that our universe is probabilistic, as framed by quantum theory. Chaos theory stuck yet another stake through the heart of predictability, telling us that even in systems that in principle are predictable, like atmospheric phenomena governing tomorrow's weather, tiny uncertainties about the present conditions expand quickly, making long-term forecasts a fruitless exercise. This is one of the reasons why weather prediction extending beyond two weeks remains a toss-up even today.

These advances create a puzzler: Could we offer mathematical proof of the fundamental unpredictability of our future actions? Given that each of our atoms and molecules is subject to quantum theory, does this make us—humans—fundamentally unpredictable from within?

Yet if we fail to discover a principle that poses strict limits on human predictability, does that mean that human behavior is in fact predictable and there is some probabilistic framework that can unveil our future actions? Is it possible that if we invest in research to uncover the laws of human actions—similar to our investment in physics with an aim to unveil the details of the subatomic world—we can make ourselves fully or at least substantially predictable? Do we accept the forbidding of Popper or chase the promise of Telegdi?

❦

So far this book has offered only a series of intriguing mysteries and questions. Dirk Brockmann found that Einstein's diffusion theory fails to explain the flight of the dollar bills, suggesting instead the existence of some invisible force that slows the flow of money. We also showed that the transmutation model fails to predict Web visitation.

So far the medieval tale we are following is equally confusing. We saw Cardinal Bakócz's failed grab for the papacy, and a year later we winced as György Székely dismembered an Ottoman warrior. The cardinal has now returned to Buda with a shiny new Crusade, the ace up his vestment sleeve, and we heard Telegdi offer dire warning against the cardinal's plan. The plots we follow diverge and meander, just like the chaotic trajectories of the weather.

The time has come to pull some of the strings together. How do our medieval characters, the Székely, the cardinal, and our prophet Telegdi, relate to one other? Why do we fail to predict even the simplest patterns that characterize human activity, from travel to Web browsing, never mind the economy? In the coming chapters we'll start to see some answers.

8

A Crusade at Last

Matthias Church, Buda, April 24, 1514, St. George's Day

 atthias Church is not the official center of the Buda Castle, though it certainly has become its spiritual core. A national shrine and a must-see in the castle district, the church sits next to the Institute of Advanced Studies, only a brief stroll from the palace where István Telegdi passionately addressed the Hungarian court hundreds of years ago. Captains once attended ceremonial Masses here and a blessing of the flag before every major military campaign. If they came back victorious, they gave thanks with the Te Deum. As they sang their praises heavenward, the standard of the defeated enemy would be hung on the church's inner walls.

In 1444 János Hunyadi, standing in this church, would have seen the Ottoman banners, captured during his triumphant winter campaign. A decade later, there also stood Giovanni da Capistrano, the aged Franciscan friar with the silver tongue, who had raised the army for Hunyadi to liberate Belgrade. So it is only appropriate that if you are willing to walk the 121 stairs down the wide staircase of the Fisherman's Bastion you will find Hunyadi standing there in bronze. His arms rest on a long sword, and he gazes across the wide Danube, as if keeping watch over distant Belgrade.

These days a series of benches takes up most of the space inside the church. They are a recent addition, however, and in 1514 the church would have been an open expanse, resembling more a busy square filled with festive, mingling townsfolk. The congregation was indeed noisy on April 24, 1514, a mixture of knights, noblemen, and prosperous merchants chatting and greeting one another.

The noise dissipated somewhat only after six acolytes in white linen albs, each with a long candlestick in hand, filed in from the narthex. After them came the deacons and priests in their ceremonial frocks. As the procession neared the altar, the acolytes moved to the side to make way for the cardinal, who, despite his advanced age, closed the procession with all the pomp and self-assurance such an occasion required.

With an unhurried, profound reverence, the cardinal placed his left hand on the altar, made the sign of the cross, and, in a voice strong and clear enough for all to hear, began the Mass, chanting, *"In nomine Patris, et Filii, et Spiritus Sancti."*

With the worshippers' amen, he continued, *"Introibo ad altare Dei."*

The congregation listened to the familiar verses and watched as the cardinal bowed toward the altar, struck his breast three times with his right fist, and intoned, *"Mea culpa, mea culpa, mea maxima culpa."*

Though the Latin rite was identical to countless recitations before it, this particular Mass was no ordinary Mass. You would have sensed as much if you had looked over next to the king's richly carved, empty throne, where ten or so men mixed, an assemblage of knights and mercenaries. The knights were landowners, and the mercenaries paid by the day. If the two started to befriend one another, there was war in the air.

One knight in a dazzlingly opulent outfit stood out in this group. Yet, if you took note of his rough hands, emerging from the exquisite gold embroidery edging the sleeves of his crimson uniform, or of the red, windburned skin only partially hidden by his exuberant mustache, you would be forgiven for suspecting that this was not his usual attire. The expensive outfit clad György Székely, captain of the Ends, whom we abandoned two months earlier as he was busy separating Ali of Epirus's arm from his body. The creaking armor, the rusting mail, and the fading crest were all gone now, replaced by a gold chain on his breast and a brilliantly polished sword with a golden pommel in the shape of a fish tail, well worth a fortune. And if you still doubted whether this sober figure was indeed our man, his brother, Gergely, was there with him as further proof to his identity—not as richly attired, but scrubbed raw for the solemn proceedings.

Killing Ali had been for György as much an act of desperation as an historic victory. Yet the final blow that brought Ali down had opened a wellspring of jubilant pride from the rough men-at-arms of the Ends. The celebration that followed had taken the elder Székely by surprise. Suddenly a hero, he was now known to everybody as the one who with a single stroke had killed the famed Ali of Epirus.

While György looted the belongings of the Ottoman captain, Ali's prized horse, skittish without the reassuring hand of his owner, galloped away. The handsome steed was later captured by Hungarian horsemen, who tarnished Székely's triumph by embroiling him in a bitter feud over the horse. He eventually gave up the horse for three hundred gold pieces, but the dispute only strengthened his resolve to leave the Ends. Now that he was possessed of sufficient money to strike out on his own, the question was only where to go. For reasons neither he nor his brother Gergely cared to share, anywhere but home sounded fine.

But just as György Székely was contemplating his next move, news came that the cardinal was on his way back from Rome. Rumor had it that he was carrying with him the crucifix, which meant that the big war, the one that would crush the Ottomans once and for all, was imminent. His experience at the Ends might prove valuable to the new campaign, he surmised, and so turned his horse toward Buda, his purse heavy with gold pieces. He stopped for a few days at the castle of Temesvár, the other major fortress in the Ends. As he lightened his pockets at the rowdy, packed taverns that lined the city's narrow streets, he had no way of knowing that the starting point for his journey was precisely the spot where it would end one day.

The Castle of Buda was roiling with crowds, its streets overflowing with knights, mercenary soldiers, and merchants. Mercurial couriers dashed through the gates, and many more arrived each day. It was spring, the trees were blossoming, and the castle's inhabitants were glad to leave behind the musty caves and small houses that had shielded them from the worst of the winter cold. As György Székely discovered the hangouts frequented by other mercenaries, he was surprised to learn that the court was already abuzz with the news of his victory over Ali, embellished with details that made him seem larger than life. Call it testosterone-fueled impulse or lunatic bravery, his feat at Belgrade had become celebrated as an epically courageous act of patriotism, an act of heroism the people had all been aching for. Thanks to this convenient exaggeration, Székely found doors opening to him that he would have never dared knock upon.

Doing little to dispel the myths, he soon found himself kneeling before the king, who praised his bravery. But his laurels weren't woven only of empty words: The spur and the sword given him by the king conferred knighthood. Land and a village of forty families, nestled somewhere between Temesvár and Belgrade, constituted a comfortable income for the rest of his life. A severed arm, dripping with blood, was added to his coat of arms, in lasting tribute to his victory at Belgrade. Double pay, a solid-gold chain, the gold-embroidered crimson tunic he now proudly wore, and the promise of another four hundred gold pieces from the royal treasury ensured György Székely and his brother the means for a worry-free life.

Despite the bounty of gifts and the ringing accolades, György failed to acclimate to his newfound fame. His quick rise in status left him feeling a charlatan, a mere nobody pretending to be somebody. His unease was further aggravated when he went to collect the four hundred promised gold pieces but was thrown out of the palace on the orders of the respectable István Telegdi. Moreover, Miklós Csáky, the powerful bishop of the episcopate where György's new village lay, opened his door to the conquering hero's knock only to reproach him for his past misdeeds. So far, his life in Buda had proved a roller-coaster ride, one day ennobled by the king, the next humiliated by the arrogant Telegdi and Csáky.

Once the Mass was concluded, the cardinal finally rose to speak, coming to the moment everybody was waiting for.

After five decades under the Ottoman Empire, Constantinople must return to its rightful stewardship, said the cardinal, emphasizing that the reclamation of the seat of Christianity was a divine mandate. He was humbled to be asked to helm the Crusade of the century, he explained eloquently, and he was fully aware of the sacrifices the holy war would make on everybody. In the pope's name, the cardinal promised forgiveness to all those who were willing to spill their blood for the cause. He then called on Count Bernard, the pope's notary, and the bishop of Pavia to read Pope Leo X's bull. After a Franciscan friar offered a rough Hungarian translation, the cardinal added that any father holding his son back from the holy fight, or any son holding back his father, would draw God's righteous anger on them both.

Then, with a spare gesture, the cardinal invited György Székely to the altar. The knight humbly obliged, his new spear loudly knocking against the cathedral's stone. The cardinal handed him a large white flag, decorated with a red velvet cross, modeled after the standard carried by the Knights Templar into the Holy Land. Blessed by the pope, the flag's destination was now Constantinople. György Székely accepted the flag and knelt, while the cardinal's tailor nimbly stitched a replica of the cross onto the red tunic.

The significance of the moment escaped no one. With this simple gesture, the cardinal had promoted a virtually unknown knight from the Ends, a Szekler from the remote Carpathian Mountains, to commander of the Crusade.

April 24, 1514, was St. George's Day, György Székely's name day, and if his parents had observed the custom of christening their child with the name of the saint on whose day he was born, it was probably his birthday as well. Could there be a more divine birthday gift? If this was not an act of God, then God never intervened in the lives of men. For a decade he had served the king, fought for and under him. Now he was being called by the cardinal to answer to an even higher authority, with that red cross patch branding him a soldier of Christ.

György had no idea why the cardinal had plucked *him* from obscurity to be saddled with this enormous responsibility. But at this moment, he really didn't care. The pope's bull promised a place in heaven to all who took up arms against the Ottomans, in victory or defeat, and he had no intention of failing. He was destined to carry the pope's white flag to the heart of the Ottoman Empire and listen to the Te Deum upon his return to Buda, as the flag of Constantinople fell from its fastener to the church wall.

But the sunburned knight from the Ends was not the only one celebrating. The cardinal was likewise jubilant about his victory in the court, having prevailed against such powerful skeptics as János Telegdi and secured the king's approval for his plans. His success in finding an appropriate captain for the campaign made his felicity complete.

The country had no shortage of experienced knights who could have led the upcoming offensive. Szapolyai, the voivode of Transylvania, had demonstrated his worth on the battlefield with every spoil and prisoner he carted back from his past campaigns against the Ottomans. But the cardinal was not fool enough to entrust the Crusade to his political rival.

Then there was István Báthory, the crippled captain of Temesvár. His military might was a family legacy: In 1475 another István Báthory, then-voivode of Transylvania, defeated the Ottoman-friendly Walachian boyars and installed Vlad the Impaler on the throne of southern Romania. Known later to the world as Count Dracula, Vlad had earned the voivode's trust during the fourteen years he was in Buda as the court's prisoner and guest. Yet another István Báthory is to be remembered as the greatest king of Poland, the only one to defeat that nation's longtime rival, the Russians. So István Báthory would have been well suited to ride at the head of a noble army. It's just that the battle-hardened aristocrat had little interest in leading a Crusade of peasants.

Just when the cardinal feared he was out of options, along came György Székely. Though unproven as captain of a large army, he was battle-tested, and peasants and mercenaries alike revered him. With no agenda of his own, he was clean and loyal. And, as everyone was aware, without the cardinal he was a nobody, a skilled mercenary, and a sword for hire.

9

Violence, Random and Otherwise

On May 31, 1991, Molly, who was preparing to go outside to her family's backyard pool, heard a knock at the door. It was just before 11:00 A.M., and she already had a full morning behind her in a trip to school to pick up her report card, marking the end of the school year. Now she was home alone in the affluent Maple Ridge suburb of Tulsa, Oklahoma, enjoying the first day of vacation, when she heard the knock. With no key to open the lock, the resourceful eleven-year-old suggested that the visitor meet her at the back door.

He was here for the yard work, the stranger told Molly. He was casually dressed, short, and slightly built, with a bright-red crop of hair and a clean-shaved but acne-scarred face.

Her parents were away, Molly explained politely, eager to conclude the interview and get into the pool.

He seemed to understand.

Before he left, however, he had a favor to ask. Did she happen to know the time?

Molly did not, but there was a clock behind her.

She turned around to read it.

In a flash, the door was open, and before she could fully comprehend what was going on, the stranger's strong arms closed around her.

~

More than two hundred miles to the south, in Texas, Tim Durham watched his father disappear behind the door of the Dallas Gun Club. After he heard the latch click, he collected the old man's protective glasses and earplugs, his .410 rifle, and a .28 with the ammo bag. He then carried his bundle to the black Lincoln Continental.

James Durham, the wealthy owner of an electronics store in Tulsa, was an avid skeet shooter, as was his son Tim. They'd driven down to Dallas together for the Pan American Skeet Shooting Competition and took the opportunity to pass a weekend with Jess and James Spoontz, longtime family friends in the area.

The elder Durham competed. Tim, a short man with a blaze of red hair, full beard, and mustache, was forbidden to use a gun, so he assisted as a referee.

Tim's twenty-ninth birthday was the following day, and he looked forward to the party the Spoontz daughters had planned. Over the past few years he'd had a few run-ins with the law—public drunkenness and a DUI, stealing a vacuum cleaner and charging seventy dollars on a buddy's father's credit card—petty offenses that were beginning to paint a worrisome pattern. To make matters worse, he'd recently sold a gun at a pawn shop, which turned out to be a probation violation. He wasn't allowed to carry a firearm. Not even to the store to sell. This latest misstep threatened jail time.

Technically even now, as he tucked his father's gear away, he might have been breaking some laws. He was in Texas, however, not Oklahoma, and as he trailed his father into the club, he felt confident that the chance of the law tracking him here was so slim it wasn't worth worrying about. So by the time Molly called 911 in Tulsa, in Dallas all the guns were safely put away, and Tim and his father had left the gun club and were on their way to the Spoontzes' for lunch.

Jean Spoontz helped Tim heat up the Olive Garden leftovers from the previous night's dinner. She saw the hungry sportsmen tuck into the familiar Italian fare before heading out with Tim's mom for a facial.

Around noon, Jess and James's daughter Cynthia swung by the house for a plate of spaghetti. By then, a four-hour drive north on U.S. 75, the calm of Maple Ridge neighborhood was shattered, rent by Tulsa police sirens.

❧

We trust no one to *always* make the right decisions, particularly those affecting our freedom. Recall Hasan's experience with the FBI: "When you're face to face with someone who has essentially the power of life and death over you, you do not behave like a rational human being." We have a jury, therefore, based on the belief that twelve people in a public setting are more likely to see the truth than any single individual in a closed interrogation room.

Indeed, even if we assume that a juror can see the truth 80 percent of the time, a juror's chances of being wrong are still at 20 percent. This is worse than your odds of getting shot in Russian roulette. So you clearly do not want to place your freedom in any one juror's hands. But with twelve jurors impaneled, each erring approximately 20 percent of the time, the likelihood that they'll *all* vote to convict you if you are innocent is only $(0.2)^{12}$, or a 0.0000004 percent chance. This means that it takes about a half billion trials for a twelve-member jury to wrongly convict an innocent defendant. This was music to Tim Durham's ears.

❧

"You think it will never happen to you. I have so much anger in me. I see no reason why someone could do something like this," Molly's mother told a reporter on July 15.

Six weeks after her eleven-year-old daughter had been brutally raped by a stranger, she had reason to despair. The detectives had exhausted all the promising leads, prompting the wealthy Oklahoma community to openly voice its anger about the unsolved crime.

Then the police had an unexpected break. A Tulsa detective's wife, who worked in the probation department, told her husband that she knew somebody who fit Molly's description of the attacker: a short, red-haired man, a fellow on probation by the name of Timothy Durham.

Molly was shown Tim's photo in a lineup, but she was not sure. They tried it again. The second time around the photo seemed a bit more familiar. "Looked like him," she said, fingering Tim's picture. And with that, in January of 1992 Tim Durham was arrested. Despite his shock, he was convinced he'd beat the rap; well over a dozen people recalled having seen him at the Dallas skeet-shooting competition, respected businessmen and church elders alike.

At the trial the defense lined up eleven witnesses who asserted that when the red-haired man raped the eleven-year-old girl in Oklahoma, the defendant was in Texas with them. They also told the jury that at the time of the attack, Tim Durham had a mustache and full beard he'd been growing out for over a year, so there was no way he was the clean-shaven man Molly recalled.

Then the prosecution was up. The assistant DA called to the stand a forensics witness who meticulously matched the red hair found at the crime scene with Tim's. Next a DNA expert testified that semen found in Molly's swimsuit matched, to some extent, Tim's genes. The marker they based the match on is common among 5 percent of the population, thus by no means placing him at the scene of the crime. Then the DA pressed on Tim's alibi, suggesting that the eleven witnesses were a bunch of old men, incapable of—or maybe even unwilling to—accurately recall the day in question. Finally, Molly took the stand and identified a clean-shaven Tim as the man who had assaulted her.

"The victim was a very impressive person," the jury's foreman told ABC News after the trial. "She was absolutely and totally positive."

So on March 13, 1993, the jury handed Timothy Edward Durham a 3,220-year prison sentence for the sexual assault of an eleven-year-old girl, the stiffest sentence for a nonmurder case in the collective memory of Tulsa County. After the verdict, Tim's mom, an elegant, white-haired, God-fearing woman, threw her Bible against the wall.

"I don't believe in the system anymore if it let something like this happen; I really don't," she said. While the jury could be misled, she could not. She had been with her son in Dallas at the time of the crime. She has never stepped foot in a church since that day.

~

Despite its occasional failures, or because of it, our juridical system should reveal a lot about human decisions and their shortcomings. Yet, as anybody who has served on a jury knows, jurors deliberate behind closed doors. Furthermore, they arrive at a verdict by consensus, which makes predicting the outcome of individual decisions largely impossible.

The Chicago Jury Project says otherwise. With their extensive surveys of ex-jurors, attorneys, and judges, the project noted that the outcome preferred by the majority of the jurors on the first ballot, before the jury discussed the case, coincided with the final verdict in 91 percent of the cases. Therefore, to calculate the outcome of a verdict, it is sufficient to consider the view of the majority. We can adjust our calculation to do just that, and now the probability that a twelve-member jury will wrongly convict an innocent defendant jumps from 0.0000004 to 0.4 percent. This change is significant, revealing that four defendants in every thousand trials is wrongfully convicted.

Some may argue that this number in fact indicates that a wrongful conviction is much more the exception than the rule. While that may be true, it's little consolation to Tim Durham.

While we can never take back Tim's suffering, we can identify a weakness in our approach. We originally started with the ad hoc assumption that jurors, on average, are correct 80 percent of the time. But could it be that they actually perform much better and that circumstances like Tim's occur much less frequently than predicted? Can we somehow determine the real likelihood that an average juror will arrive at the right conclusion?

Since it can never be known for certain who is guilty and who is innocent, the task seems hopeless. But as we've learned from a nineteenth-century French mathematician, some things are not nearly as difficult as they initially appear.

~

Siméon-Denis Poisson recalled once how, as a toddler, he was found by his father, suspended by a cord. The culprit was the maid, who insisted that she had only taken a necessary precaution to protect the small boy from bugs and animals that roamed the floor. The truth was that she confined the child so that she could go about her own business.

Despite his early brush with the mathematics of the pendulum, the young Siméon-Denis was sent by his family to study medicine in Fontainebleau. Scarred by the death of his first patient, he abruptly quit his studies and fled home.

While he was away at school, Siméon-Denis's father had been elected mayor of Pithiviers, a status that demanded subscriptions to a number of impressive periodicals, including the *Journal of the Polytechnical School*. With little to occupy him in the countryside, Poisson began paging through the journal and awakened an unexpected predilection for the math puzzles inside, cracking them one by one. His family reconvened to once again discuss his future and sent the bored youngster back to Fontainebleau, to study mathematics. This time he did not quit (it's harder to kill people with equations) but entered the Polytechnique in Paris and seven years later succeeded the legendary Fourier as a professor there.

Today Poisson is remembered for a wide spectrum of scientific discoveries—like Poisson's integral, Poisson's equation in potential theory, Poisson's ratio in elasticity, and Poisson's constant in electricity, for example. His name is engraved on the Eiffel Tower, and a heavily eroded crater in the southern highlands of the Moon's near side, east of Aliacensis and northwest of Gemma Frisius, bears his name. He wrote more than 350 papers, an output considered just as remarkable then as it is today. And that was before the word processor.

His best-known result was published in 1837, only three years before his death, under the title *Researches on the Probability of Criminal and Civil Verdicts*. Its subject, the design of a good jury system, holds little interest for mathematicians today. Yet the importance of the work cannot be overstated, as it established the foundational theory for statistics, a set of mathematical tools used in most research fields today.

In this seminal work Poisson showed that the likelihood that a typical juror will err is calculable. All we need is the number of trials and convictions that take place in a given year. Such information was regularly collected by the French Ministry of Justice, indicating that of the 6,652 defendants accused of crime in 1825 only 60 percent were convicted. Using these numbers and a formula he derived, Poisson determined that individual jurors make the right decision on average only 75 percent of the time. What is particularly fascinating about this result is not the percentage he calculated but his ability to get into the heads of individual jurors, determining their rate of correctitude.

Using the statistics collected by the Chicago Jury Study, Alan E. Gelfand and Herbert Solomon concluded that U.S. jurors perform slightly better: They make the right call a good 90 percent of the time. They apparently approach their duty with more gravity and wisdom than did their nineteenth-century French forebears. But does this really quash any lingering doubts we may entertain regarding our current jury system?

For Tim Durham, it is unlikely. But for the rest of us, things do get slightly better. Indeed, if individual jurors are right 90 percent of the time, an innocent defendant will be convicted only 0.005 percent of the time. This is almost a hundred times better than the expected wrongful conviction rate if the individual jurors err 20 percent of the time. Given that in 1990 there were 1,993,880 convictions for index crimes, including murder, manslaughter, rape, assault, robbery, theft, and arson, according to this prediction only about 40 of them should have been wrongful convictions. This may not sound like a lot, unless you happen to be one of the unlucky 40.

❦

There is a philosophically deep hypothesis underlying Poisson's work: He made the simple but loaded assumption that human behavior is random. You may be the smartest person on earth or the most obtuse, lawyer or criminal, skeptic or believer, but once you sit on the jury bench, for all practical purposes all we know about you is that on average you recognize the truth only 90 percent of the time. With that, Poisson equated *unpredictability* with *randomness*. He then proceeded to show that once we accept that human behavior is random, it suddenly becomes predictable.

This is somewhat of a paradox: If unpredictability implies randomness, how can randomness mean predictability? The answer is simple: Poisson's predictions come in a quite different guise from the predictability we seek in our daily life. It is not akin to the prophecy of István Telegdi, projecting the future of the papal Crusades. Rather, it is much more similar to Einstein's derivation of the laws of atomic motion. Indeed, Einstein knew that it was impossible to predict the trajectory of any single atom. He showed, instead, that if we accept that atomic motion is truly random, then an atom's typical distance from its release point will follow the laws of diffusion.

In a similar fashion, Poisson never aspired to predict whether as a jury you are right or wrong. He assumed, instead, that each juror votes as if tossing dice: Most of the time they are right, but occasionally they make mistakes, and we never know for sure *when* a juror is right or wrong. Equipped with this assumption, from the dry statistics of the conviction rates Poisson was able to determine the reliability of the jury system as a whole.

To better understand how Poisson's predictions work, consider my phone records, which indicate that on average I make about twelve calls each day—or about a call about every two hours. That does not tell you, however, *when* I'll call next. But if you assume that my call pattern is random, you suddenly understand quite a bit about my communications. Using Poisson's formula, you can determine the likelihood that I will not make a call over the next hour (the chance for that is 60 percent—that is, quite likely) or that I will make five calls in the same period (0.02 percent—that is, unlikely). His formula also reveals the likelihood that I won't make any calls all day (0.001 percent—that is, very unlikely).

Though a far cry from the presage of an oracle, such predictions can be extremely valuable. Consider, for example, an engineer at a phone consortium whose job it is to decide the capacity of the cell-phone tower to be erected in your neighborhood. If he sets the capacity too low, many calls will be dropped, angering the consumers and his boss. Too much capacity wastes the company's resources, however, decidedly angering his boss. If the engineer knew precisely when you and everybody else in your neighborhood planned to use their cell phones, he could precount the calls to be made at peak times and size the tower to handle the load at its heaviest.

No engineer is privy to your future calls, however. Rather, he knows that the typical cell-phone customer makes an average of three calls per day. He also assumes that everybody's calling pattern is random, so, using Poisson's formula, he can predict the number of people planning to use their phones at any given time. He will then add enough capacity such that no more than three out of a hundred calls on average will be dropped, ensuring his company meets the industry's benchmark for "spotless" mobile service.

Today, whenever a series of events is beyond comprehension, we say that they are random. As we shall see, such apparently random acts drive historical dramas as well, providing twists and turns that are best understood in terms of rolling dice. But what does being random really mean?

To get an idea, roll a die, and jot down a dash (|) on a piece of paper every time you get a six. Place a dot (.) whenever you get anything else—that is, a one, two, three, four, or five. I just did this exercise myself, and after about four hundred throws, the sequence I got looked like this:

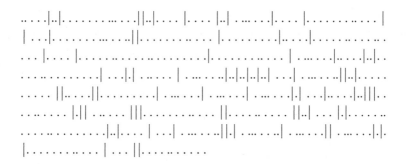

While the outcome of each throw is utterly unpredictable, the sequence itself has some uniformity to it. That is, most of the sixes fall within roughly five to seven throws of each other, and our chances of not getting a six for a hundred throws—which would look like a really long gap between two dashes—is negligible. Indeed, Poisson's formula tells us that this happens only once in about a hundred million rolls of the die. Similarly, we would need about a hundred million throws to yield ten dashes in a row, representing a lucky sequence of ten sixes. So, while the outcome of the next throw is always a mystery, true randomness has some magic uniformity to it.

Despite this apparent uniformity, there is nothing more random than a Poisson process, which is nothing but a sequence of truly random events. Therefore, deviations from Poisson's predictions are often taken as a signature of some hidden order. They offer evidence of a deeper law or pattern that remains to be discovered.

To be sure, many phenomena we encounter are obviously not random, like the motion of the planets and the reassuringly repetitive cycle of day turning to night turning to day again. Other phenomena, like the weather, appear to be driven solely by chance. As Richardson so eloquently pointed out, the atmosphere is governed by laws and equations, which today are successfully exploited by meteorologists everywhere. Many of these events—from solar eclipses to floods and droughts—were once considered the mysterious provenance of gods and spirits. Today they are predictable, however, telling us that deviations from randomness are often signatures of fundamental laws yet to be discovered.

～

Tim Durham's family convinced Barry Sheck to take over Tim's defense. Sheck, the DNA expert on the defense team that won O. J. Simpson's acquittal in 1995, cofounded the Innocence Project, which uses DNA-based techniques to exonerate the wrongfully convicted. A new DNA analysis, relying on technologies unavailable at the time of the first trial, unequivocally excluded Tim as the rapist.

Tim had already served five of the 3,220 years of his sentence before he was released. There is a hierarchy in prison, and child molesters are squarely at the bottom of it. After his incarceration for Molly's rape, Tim was beaten so brutally that his ribs cracked. Once free, he learned that another young girl in Molly's neighborhood had been similarly victimized a month before Molly's assault. The descriptions of the two girls' attackers were remarkably similar. Tim also heard about a man, Jess Garrison, around five feet tall and with reddish-brown hair, who had vanished from the area before Tim's arrest in 1992. Garrison hanged himself on December 18, 1991, two years before Tim was convicted. His DNA was never tested.

I wish I could say that Poisson's work has led to the best jury system possible. But it hasn't. The evolution of the jury continues to be driven less by science than by legal and political argumentation. Yet Poisson's work did answer an important question: Is there such a thing as a perfect jury? Sadly, the answer is no.

But can we at least improve the odds of getting the verdict right? Poisson's math does offer us some guidance on that score: The more jurors on the jury, the smaller the chance that they'll collectively err. That is why it's so surprising that in the 1970s the U.S. Supreme Court decided that certain juries may be comprised of as few as six members.

The proponents for the change had argued that reducing jury size resulted in considerable savings, as fewer people would be required at the courthouse. Smaller juries also meant fewer hung juries and reductions in deliberation time, as six jurors agree faster than do twelve. So Salim Hamdan, Osama bin Laden's driver and bodyguard, the first detainee to face a war-crimes trial in Guantánamo Bay, was convicted by an anemic jury comprised of six military officers. Granted, he may have been far from innocent, but with a six-member jury the probability that he was wrongfully convicted is twenty-five times greater than if his case had been tried before a jury of twelve.

Though it has not reformed the juridical system, Poisson's math remains influential. Indeed, our actions can easily appear random to those who do not know us. I do not know when you normally wake up, when you plan to send your next e-mail, when you will make your next phone call, when you will have sex, and when you will be exposed to influenza. But each of these individual actions has consequences, affecting insurance companies and telephone conglomerates, hospitals and chain restaurants, customer-service centers and stock brokerages. And whenever they are in doubt as to how we behave, scientists and engineers turn to Poisson for aid.

In 1915 it was discovered that the number of accidents follows Poisson's predictions, turning his math into the bedrock of the insurance industry. Internet routers today are designed assuming that traffic, driven by our haphazard browsing and communication patterns, follows a random Poisson process. Poisson's formula is used to describe the number of deaths from an epidemic or the number of instances of enteric fever in a household. Along the way, scientists have tacitly agreed to subscribe to the fundamental paradigm that continues to drive the science of human activity: For all practical purposes, our behavior is random. Unpredictable. Episodic. Indeterminable. Unforeseeable. Irregular.

There is only one problem with this assumption. It is simply wrong.

10

An Unforeseen Massacre

Apátfalva, the morning of May 23, 1514, a month
after the Mass that launched the Crusade

 hose who dared to travel despite the unrest that plagued the country during the Crusade would surely have presumed that any gathering of two thousand along the Maros River had to be part of the army heading toward Constantinople. Yet it is hard to imagine a more pitiful scene than the one that would have greeted travelers at the ford at Apátfalva on the twenty-third of May in 1514.

Forget the white tents with tall flagpoles and colorful banners whipping in the breeze that dominate the skyline of a medieval military camp in our Disneyfied imagination. Picture instead makeshift shelters constructed of filthy, stretched mats and canvas; a few dingy rags hung on poles stuck in the mud-trampled ground.

Forget knights ensconced in shiny, engraved armor with colorful coats of arms. Imagine instead peasants, haiduks, shepherds, smiths, and merchants clad in worn coarse wool or linen tunics, reeking of campfire smoke.

If you ignored for a moment the patchwork armor some of the men tied to their bodies and their notched, dull axes and jagged lines of stitches, this gathering of about two thousand, chatting, sleeping, and eating around campfires, could be easily mistaken for a bustling fair rather than a military camp.

Only the red cross stitched to select backs indicated their mission: They were crusaders, this sorry lot, the volunteers who had responded to the cardinal's call. They had just finished crossing the Maros River and had no intention of staying here—the road across the forest would lead them soon to Belgrade, the last bastion of Christian Europe. Beyond that, Ottomans guarded the long road to Constantinople.

After the cardinal swore him in, György Székely crossed the Danube to Rákos, a sleepy village next to Pest where volunteers had been gathering in the past few days. His arrival was not the triumphant entry of a captain about to take on the Ottoman Empire. Rather, the ragtag army awaiting him consisted of fewer than three hundred peasants, insufficient to conquer a modestly fortified village.

Most disappointing, there wasn't any armament to speak of. Some men proudly carried long sticks sharpened at the end as makeshift lances; others held timber cutting axes with long handles instead of maces and real battle-axes. Most, however, were armed only with a scythe, which only a few days before had been sharpened to bring in the harvest and now was extended upward from the staff. Only a few battle-weary mercenaries wearing odd pieces of armor evoked for György Székely the professional units he was accustomed to.

György could not believe that this sorry assemblage dared call itself an army. Yet, the cardinal had pledged additional volunteers, food, and armament, so the captain refused to be discouraged. He plunged into training his troops instead, teaching the peasants the basics of warfare—how to use a firearm, avoid a sword, kill with an ax or a spear. He was everywhere all the time, turning the haiduks with horses into a makeshift cavalry to stage mock attacks against his newly minted infantry or demonstrating the use of long halberds to counter the reach of long lances.

True to his promise, the cardinal asked that the Franciscans spread the news of the Crusades, as their order had to such great effect in János

Hunyadi's hour of need, five decades earlier. With a network of monasteries spanning the country, often themselves recruited from the local peasantry, the monks were indeed the best emissaries of the crucifix. It was a testament to their influence that, much to the astonishment of just about everyone concerned, the army of three hundred near Buda had soon swelled to fifteen thousand, with a rumored forty thousand additional troops gathering across the land.

Given the exceptional success of the recruitment, a mere two weeks after the Mass, György Székely was ordered to start the campaign. On May 10 his vast army had swung into motion, heading east, straight toward Transylvania. As the enormous snake of mismatched soldiers wended across the Hungarian lowland, it lengthened. Knights and peasants alike continued to flock from surrounding lands. By mid-May, when Székely arrived at Gyula, about fifty miles northeast of Apátfalva, he commanded more than thirty thousand troops.

Medieval campaigns normally started around midsummer, after the crops had been collected from the fields. The cardinal did not have the patience for delay, however, and so he launched the draft in April, much earlier than was customary. He gave the crusaders gold to pay for food and armament and opened his vast granaries and donated herds to feed them. These resources could easily have provided for several thousand mercenaries. But for forty thousand or so hungry volunteers, such quantities of materiel were laughably little.

Sensitive to potential food shortages, György announced severe punishments for anyone caught robbing civilians. But discipline could not feed an army, and each unit was forced to scrounge for food however they could find it. In and of itself, this was not unusual—the local populations always fed medieval armies, who took food by force if necessary. But this time, things were different: In the past, with an army comprised of knights and noblemen, only the peasants suffered the campaign's consequences; the army of serfs had no qualms, however, about robbing the farms and castles of the landowners, even spilling noble blood if faced with resistance.

There was another problem, just as systemic as the food shortage: Landowners all over the country saw their fields emptying of serfs just as the summer work was about to begin. Desperate for working hands, some locked up their men and even executed those caught trying to join the Crusades. Rumors of imprisoned and tortured would-be-crusader families spread quickly through the camps, prompting many resentful peasants to see the nobility not as their ally but, to the contrary, hostile to their cause.

As news of conflicts between landowners and crusaders reached the court, the cardinal was pressured to rein in the chaos. He finally relented and on May 14 sent a missive to György Székely, ordering him to turn away any additional volunteers. Yet thousands continued to risk their lives to reach the camps, making it very nearly impossible to adhere to the cardinal's command.

György Székely concluded that the letter did not reflect the cardinal's real desires; rather, political forces opposed to the Crusades were likely forcing his hands.

"I am not a kid or a nut to be deceived," records the *Chronicle of Sz-erémi* of his reaction to the cardinal's letter, adding, "I warn you on God and the Holy Cross!" With that, he ordered the priests to continue consecrating the volunteers.

The letter did have some impact, however: György abruptly changed course, abandoning his march toward Transylvania, which only would have made sense had he been seeking further recruits. Instead, he made a ninety-degree turn to head due south, following the shortest path to Belgrade. To get there he had to cross the wide Maros River by the Apát-falva ferry. He sent an outpost of about two thousand to the ford to prepare it for the arrival of the main army.

With the closest Ottoman fortress at least two weeks away, everybody at the outpost was hopeful of a good night's sleep after the long march. And so a general sense of drowsy apathy settled on the camp.

Given the lack of imminent danger, most of them were more surprised than frightened when their afternoon rest was interrupted by a dull but escalating clamor. It was the unmistakable sound of an approaching cavalry. Though it was yet invisible, the drumming of the horses' hooves percussed through the camp. Some of the soldiers instinctively reached for their arms; most, however, curiously drifted toward the edge of the camp so as to be among the first to spot the approaching men.

It took minutes for the turbulent sea of gray steel and mail of several hundred mounted knights to emerge from the nearby forest. The humanist Taurinus, in an epos published four years later, vividly captured the man leading them:

> *Shiny helmet covered his skull,*
> *His shin in armor and both hands covered in gilded*
> *Fluted gauntlets, on his waist*
> *two aureous basket-hilted swords with ivory hilts.*

Those familiar with his red-and-white coat of arms soon recognized István Báthory, captain of the Temesvár fortress and military commander of southeast Hungary. Accompanying him were the forces of Miklós Csáky, the bishop of the nearby Csanád episcopate who had castigated György Székely in Buda some months earlier. Countless mounted noblemen followed, as law and custom required that they fight each time their country called on them. They were the expensive mounted cavalry that dominated battlefields with their speed and deadly power, a force the crusader army, comprised chiefly of foot soldiers, had most sorely lacked.

György Székely was not overly concerned about the lack of cavalry, as his peasants were meant to be only the middle flank of a three-front attack on the Ottoman army. Indeed, despite Telegdi's concerns, the king's council had not only approved the cardinal's plan to call on the peasants but also mobilized the country's regular forces. Péter Beriszló, the Serbian ban, had hired new mercenaries to swell his forces. Despite his opposition to the entire Crusade, Szapolyai, the Transylvanian voivode, could ill afford to be excluded from an all-out war against the Turks. Hence, he, too, had signaled his willingness to contribute his Transylvanian cavalry to the common cause. And so, as Székely's small group of outriders peered at the approaching mounted troopers, they saw that Báthory's southern forces were already assembled, ready to join them on their march to Constantinople.

The huge warhorses kicked up choking clouds of dust as they thundered near, a fearsome sight, their nostrils flaring with the exertion. At that moment many of the peasants first began to understand what it might mean to wage war. But it was only when the cavalry had drawn quite near that the crusaders noted in horror that the knights' visors were shut, their lances pointed, and swords drawn. Grimly purposeful, they uttered no battle cries or guttural screams, but spurred their powerful mounts, panting, glistening with sweat, ever faster into the heart of the crusaders' camp.

Confusion and panic drove the peasants to instinctively swarm together, desperately searching to recall the training they had received over the past few weeks. Yet the superior armament of the approaching unit left little doubt as to the probable outcome of the imminent clash. Before the crusaders could get their act together the cavalry was cutting through their ranks, leaving in their wake rows of wounded, dying, and dead.

In all of this incredible account, the most surprising is that the peasants, most of whom had never seen a battle before, managed to beat back the first attack. Their victory was short-lived, however, as the mounted knights, calm and composed, turned around, regrouped, and renewed their assault. Caught between the swords of the experienced mercenaries and the swells of the Maros River, the peasants began a chaotic retreat, wading into the improbable protection of the sluggish river. It was a false refuge. Those whose armor did not pull them down to suffocate on the muddy riverbed were soon, mercilessly, cut down by the knights and mercenaries.

What was this? you may ask, and surely such thoughts raced through the peasants' minds as they ran for their lives. The last we knew, the serfs and knights were allied. They had all signed up to face the Ottomans as a united front, and all were the loyal subjects to the crown. And why again would the aristocrats cut down their workforce? What were the knights trying to prove in their cowardly attack on the meager troops at the unprepared outpost, given that the full force of György Székely's army, all thirty thousand armed men, was only one day behind?

From any perspective, the scene made no sense, a seemingly random turn of history, with motives perhaps as obscure then as it is, frankly, to us today. But was it random, indeed? And how do we distinguish random from deliberate in matters of warfare, the deadly games men play when they carry lethal weapons?

11

Deadly Quarrels and the Power Laws

By the time Hitler emerged onto the political scene, Lewis Fry Richardson had finished with ambulance driving and, well beyond the draft age, settled into a cozy job as principal at Paisley Technical College. Increasingly disturbed by Germany's military ambitions, in 1940 Richardson made an unexpected move and resigned his position at the school, choosing to live out the rest of his life on a meager pension. He wanted to focus on a problem he thought worth the financial sacrifice: discovering the laws that govern wars.

Richardson was convinced that if he could understand the mechanisms that drive conflicts, he could avert further bloodshed. His inquiry resulted in *Statistics of Deadly Quarrels*, his second book, which, like his tome on weather prediction, was densely packed with tables, formulas, and equations. This was about wars, however, and a highly unusual take on the subject at that. "I am sure that it is mathematically sound," reflected one of his contemporaries, "although the uses to which mathematics is put are most unusual—it is this which has led to his being called a crank."

Sure enough, Richardson never found a publishing house willing to risk its money on the unorthodox work. When it was finally published seven years after Richardson's death, it proved his project to predict wars to have been on a par with his weather forecast: a complete failure.

Richardson compiled a detailed catalog of all known wars and conflicts between 1820 and 1949, meticulously recording the pertinent details, from the number of casualties to the religion of the combatants. His goal was to quantify the role of the traditional causes of wars identified by the experts. Are countries that greatly differ economically more likely to fight one another? Are groups with a common language less likely to go to battle? Could an arms race be indicative of imminent hostilities? Would a common hatred for a third party reduce the chance of two groups going to war? Such were some of the basic hypotheses of war theory, and Richardson aimed to probe them with his far-reaching mathematics.

In the end, he answered no to each question, having proved that they account for nothing more than myth and misconception. He then concluded his work with brutal sincerity, saying, "Hardly anything in the way of new knowledge as to the 'causes' of wars has emerged from this monumental analysis." As far as the data were concerned, wars and battles—like the one we witnessed in the previous chapter—were simply random accidents.

≈

I am inclined to think that each e-mail I send has some purpose, and therefore its timing is never random. I must admit, however, that to an outside observer, not privy to my motivations, the sequence of e-mails I sent on Friday, August 18, 2006, may appear to have been governed by chance. It started with an e-mail at 8:49 A.M., followed by thirty-one more, sent at 9:46, 9:49, 10:38, 11:49, 11:49, 11:53, 11:57, 1:46 P.M., 1:47, 1:48, 1:59, 2:41, 2:56, 2:58, 2:59, 3:18, 3:20, 3:30, 3:53, 3:58, 4:05, 4:05, 4:07, 4:37, 4:42, 4:52, 5:05, 5:06, 6:16, 6:16, and 6:19 P.M., a sequence of time stamps that could easily have been produced by a random-number generator. If so, my e-mail pattern should sit comfortably in Poisson's universe, the mathematics of random events built on the assumption that everything we do is driven by chance. But is this sequence indeed random?

The first five e-mails, starting with a reply to a postdoctoral associate at

8:49 A.M. and ending with an 11:49 message that carried the fruit of my morning's work, is quite consistent with a Poisson process: five e-mails in three hours, the average time between them roughly forty-five minutes. Then, in an eight-minute interval, between 11:49 and 11:57, I fired out four more e-mails, all related to my morning work. According to Poisson's formula the likelihood of such a rapid succession of messages in a random stream of e-mails is 0.000035—that is, it should occur only once every five months. Perhaps this mid-August Friday was not that ordinary after all.

After 11:57 things returned to normal, as a bike ride to the university and lunch kept me away from my computer for a while. Yet, starting at 2:41 P.M., I was off the charts again, sending eleven e-mails in the next seventy-one minutes. Again, as things go, this is by no means remarkable. If we accept, however, that my activity pattern is random, Poisson's theory predicts that such a rapid-fire e-mail stream should happen only once in 10^{26} years. Given the estimated 10^{10}-year lifetime of our universe, I evidently achieved something remarkable that day.

In reality, that Friday was by no means special. To the contrary, it was so ordinary that if the computer had not kept a record of my e-mail pattern, I would surely not remember at all what I did that day. And the pattern was by no means unique, for if I inspect any other day's e-mail log, I encounter a strikingly similar picture. So what's the fuss about?

The problem is that for human activity to be random, it must be uniformly so, translating into a stream of comfortably paced e-mails. But that's not what happens with my correspondence. Instead, on any given day long e-silences are followed by short periods during which I send a large number of messages. The truth is that no matter which day I choose to check, my sequence of e-mails does not look random. Ever. Instead, it is full of *bursts*.

~

In the late 1980s, when, as an undergraduate at the University of Bucharest, I started to read about chaos theory, one of my heroes became the Swiss mathematical physicist Jean-Pierre Eckmann. During his pioneering career, Eckmann brought order to chaos. He helped turn the butterfly effect into a rigorous field packed with theorems only fully appreciated by a handful of specialists who grasp his highly sophisticated mathematical language. Then, around the year 2000, after a career spent writing papers with titles like "Ergodic Theory of Chaos and Strange Attractors," Eckmann's work took an unexpected turn.

"Someone asked me whether one could find what 'revisionists' [Holocaust deniers] wrote," he explained. "I did not feel like reading their nonsense but found it challenging to scan the Web for their pages."

So he built a search engine that automatically searches the Internet for pages with revisionist content. As Eckmann's pseudo-Google filled his computer with anti-Semitic rants, he noticed an interesting pattern: Revisionists frequently linked to one another, forming an easily identifiable community on the Web. There was one blatant exception—the homepage of an aerial observer from Australia, to which many of the revisionist sites linked, but whose content had apparently little to do with the Holocaust.

"I got worried about the method," recalls Eckmann, "but upon inspecting the page by hand, I found that the site was cited because it said that no smoke was seen in an aerial photograph of Auschwitz." Revisionists take that as proof that no one was burned there.

The mathematical approach Eckmann had mastered over the previous thirty years, heavily punctuated with proofs and theorems, proved about as useful to his new quest as butterfly nets would be in harnessing an approaching tornado. But Eckmann's new work was no mere diversion, and two years later he published another paper on the online universe, this time charting e-mail communications. He first assembled the e-mail record of thousands of students, faculty, and administrators from a university he refuses to name. In a privacy-obsessed world, gathering such data was not an easy task, so he understandably cannot reveal his source. But during a visit to my research group he generously shared with us the anonymized version of the e-mail records. And when I finally analyzed the data in the inspiring environment of the Buda Castle in the spring of 2004, the verdict was clear: Nobody's e-mails followed the coin-flip-driven dull, uniform rhythm that characterizes a random Poisson process. Instead, each user's e-mail pattern was similar to mine—it was bursty, like the stormy late-summer weather, a thunder of e-mails followed by long periods of silence. As we saw in the previous chapters, deviations from a purely random pattern must not be ignored, as they may reveal overreaching laws of society and nature. Which happened to be the case here as well.

ᴭ

In *Statistics of Deadly Quarrels,* Richardson's opus on war and peace, he did find a striking deviation from randomness: the magnitude of the conflicts. Some wars had millions of casualties, others only a few dozen, huge differences that prompted him to measure a conflict's magnitude using the ten-based logarithm of the total number of deaths. In his classification, the February 28, 1514, skirmish at Belgrade between the Hungarians and the Ottomans was of magnitude zero, as we know of only a single casualty, Ali of Epirus. A fight with ten casualties was of magnitude one, and a conflict that left a hundred dead, magnitude two. The massacre we witnessed in the previous chapter, therefore, in which more than a thousand peasants perished between the advancing cavalry and the swirling river, was of magnitude three.

If wars were truly random, most conflicts should have a comparable number of casualties. Richardson found, however, that of the 282 wars between 1820 and 1949, 188 were relatively small, of magnitude three or less, or only a few thousand deaths. There were fewer hostilities numbering about ten thousand casualties—namely 63 conflicts of magnitude four. He found, however, 5 conflicts of magnitude six and 2 of magnitude seven, wars that each took the lives of tens of millions.

You'll easily guess the magnitude-seven wars—the two world wars. The six conflicts claiming about a million or more casualties each are much less obvious. In their order of magnitude, they were the Taiping Rebellion (1851–1864), the Spanish Civil War (1936–1939), the First Chinese Communist War (1927–1936), the Great War in La Plata (1865–1870), the North American Civil War (1861–1865), and the sequel to the Bolshevik Revolution (1918–1920).

Upon inspection, Richardson realized that the numbers of casualties followed a simple mathematical law, which he condensed in a simple phrase: "the fewer, the larger." That is, the vast majority of conflicts were minor skirmishes, leaving at most a few hundred dead. But grand clashes, with enormous numbers of casualties, were less and less likely to appear in his database.

Richardson was not the first to discover this pattern. Indeed, the nineteenth-century economist Vilfredo Pareto found that while the vast majority of people are poor, a few individuals amass outlandish wealth. The existence of the rich isn't surprising, as even if wealth is acquired randomly some will be richer than others. Pareto found, however, that the rich were far richer than a random distribution of wealth could ever explain. Richardson's and Pareto's work showed that wars and income follow what we call a *power law*. Specifically, many small events coexist alongside a few extraordinarily large ones.* It meant that for every world war and every Gates or Rockefeller there would be countless small conflicts and millions of poor.

I had a close encounter with power laws in 1999 while studying the popularity of Web pages. My research group and I found that while most Web sites are completely unknown, a few hubs, like Google, Amazon, and Yahoo!, have amassed millions of links. We named such hub-dominated networks *scale-free*, and I spent much of the following decade studying their role in complex systems, from the cell to the Internet.

My experience suggests that Richardson's "the fewer, the larger" maxim is misleading, giving the impression that a key feature of a power law is the paucity of big events—like world wars, the superrich, or hubs on the World Wide Web. We should somehow expect them less. To the contrary, we can, in fact, count on their occurrence. It is Poisson's world that forbids these outliers. In a random world, Google and Yahoo! could not attract millions of links, Bill Gates could not amass billions of dollars, and no war could claim the lives of millions. Yet, the world is just not like that. The very essence of a power-law distribution is that it naturally predicts such rare events, telling us that we will always have a few data points that grossly deviate from the average. In other words, when power laws are present, we can always count on outliers.

---

* In mathematical terms, Richardson discovered that the number of wars of size s follow the power-law relationship $P(s) \sim s^{-\gamma}$, where γ is called the *scaling exponent*.

≈

The conclusion we reached from Eckmann's data was simple: None of the e-mail users in his database corresponded randomly. Instead, their usages display the same pattern: short periods of intense e-mail activity followed by long periods, often days, of no e-mails. This, of course, makes lots of sense. We attend meetings, watch movies, go out, eat, sleep, and participate in a wide array of activities that keep us away from our computer. Once we finally check our e-mail, we tend to send several messages in a short time frame, generating a burst in our e-mail pattern. Then other tasks take us once again away from the computer, marking the beginning of another break in our stream of e-mails.

Given this rhythm our lives take, bursts in activities are by no means exceptional. That said, the lifestyle you pursue versus that which I pursue is sufficiently different, such that we would not expect that our e-mail patterns would show any similarities. Some people send only a few e-mails a week; others close to a hundred each day; some peek at their mail only once a day. Still others practically sleep with their computers. This is why it was surprising that, when it comes to e-mail, everybody appears to follow *exactly* the same pattern. Indeed, when we measured the time between the consecutive e-mails sent by the same person, no one obeyed the familiar Poisson distribution. Instead, no matter the person, power laws greeted us.

Once power laws are present, bursts are unavoidable. Indeed, a power law predicts that most e-mails are sent within a few minutes of one other, appearing as a burst of activity in our e-mail pattern. But the power law also foresees hours or even days of e-mail silence—comparable to the rare epic wars in Richardson's data set or the superrich in Pareto's analysis. In the end, the patterns of our e-mailing follow an inner harmony, short and long delays mixing into a precise law, a law that you probably never suspected you were subject to, that you never made an effort to obey, and that you most likely never even knew existed in the first place.

≈

So what? Chances are that your life does not revolve around e-mail, and even if it does, why should you care about the mathematical law that your messages follow? You didn't care when we thought it was random, so why worry now that it appears to be nonrandom?

Counterintuitively, the main reason this burstiness fascinates is precisely because it is *not* unique to our e-mail pattern. For example, if we visit our favorite Web site, we normally click around for a few minutes, reading a few articles before moving on. It is hard to believe that this haphazard pattern, driven by our mood and availability, should follow any law. But it does. When my research group measured the time spanned between consecutive clicks by the same user on a Web page, a power law greeted us once again.

Prompted by the unexpected similarity between our e-mail and Web-browsing habits, I started to search for other data that captures various aspects of human activity. Soon I discovered that Maya Paczusky, a physicist at the Imperial College in London, and her student, Uli Harder, had examined people sending jobs to a printer, studying the time lapses. A bursty pattern greeted them as well: We print many documents over short time intervals and then move on to other things for a while.

The Hesburgh Library at the University of Notre Dame graciously provided my team with records detailing the times that students and faculty checked out books. Unlike our counterparts at the FBI, we didn't care who read what; we merely wanted to know *when* each patron visited the library. Burstiness was present here as well—a typical library user checked out several books within hours of one another, perhaps preparing for a class or reading up on a research assignment. Then came long periods during which that same person appeared to forget about the library altogether.

Calling patterns were similar. Most calls we make come in short time intervals, surrounded by long breaks with no calls. And our globetrotting friend from Chapter 1, Hasan Elahi, gave us the time stamps from the thousands of pictures he took as he documented his own whereabouts. There, too, a power law greeted us: Hasan tended to take numerous snapshots over brief periods of time. He then acted as if he'd lost his camera, snapping no picture for hours or even days on end. Naturally, this might make the FBI suspicious—where was he during those periods?

No matter what human activity we examined, the same bursty pattern greeted us: long periods of rest followed by short periods of intense activity, like the pleasing sound of violins interrupted by the violent roar of drums in a Beethoven masterpiece. Bursts were everywhere in nature, from the edits individuals published at Wikipedia.org, to the trades made by currency brokers; from the sleeping patterns of humans and animals, to the tiny moves the juggler makes to keep his sticks in the air. It was no longer e-mail or Web browsing we were studying; we were witnessing something deeply linked to human activity, something announcing loud and clear that none of us behaves randomly, ever. In and of itself, this was not surprising, as no one really believes he or she is ruled by chance. We all have free will, which complicates everything, including our e-mail, printing, and Web-browsing activity. Yet no matter what we did, we unconsciously followed the same law—a power law. Conceptually simple, yet rather surprising.

❧

Any law or pattern is important only if it makes clear things previously unexplained. Indeed, Newton's law of gravitation would have had much less of an impact if it had not predicted with striking accuracy the trajectory of planets, rockets, and satellites. Burstiness has similar predictive power.

Do you remember when we conjectured that the life of a news item on the Web is about thirty-six minutes? But we soon after discovered that, in fact, most documents on the Web have a much longer life—thirty-six hours, to be exact. We arrived at the thirty-six-minute prediction by first assuming that our clicking was random, just like the thorium atoms, when they inexplicably turned into radium. Burstiness changes everything, however. It means that the approximately twenty times a typical user daily clicked on the site were not spread uniformly during the day but were concentrated in a few distinct bursts.

Once we allowed for bursty visitation, our same theory suddenly provided the correct result: thirty-six hours, not minutes. Indeed, we don't click once every hour or so on our favorite Web site. Rather, once we visit, we unleash a barrage of clicks. It may take hours or days until we return to the site, and thus it takes much longer than thirty-six minutes for us to notice any updates, fresh stories that would inspire another burst of clicks.

Dirk Brockmann, our dollar-bill tracker, also started from the assumption that the times at which dollar bills are recorded at Wheres George.com follow a random but uniform Poisson process. His was a reasonable hypothesis, given the unpredictability of the Georgers' travel. Upon closer inspection, Dirk found that the times between consecutive banknote sightings followed a bursty pattern. A particular dollar bill was frequently spotted for some period, and then it vanished, taking months to reemerge once again.

Take, for example, the record-holding dollar bill we discussed back in Chapter 3: It was spotted twice in July 2002 and then vanished for half a year, only to reemerge seven times in the following three months before disappearing once again for several months. Nobody knows where the bill was during those long breaks—sitting in the glove compartment of a car? Hiding in the pocket of a pair of retired jeans in a teenager's closet? Either way, one thing is clear: The time between any two sightings was not random but followed a power law. And once Dirk incorporated this bursty pattern into Einstein's diffusion theory, the predictions for the dollar bill motion were significantly slower, in full agreement with the mysteriously sluggish spread of the money.

As a pacifist Quaker, Richardson believed that no resources should be spared in attempting to avoid the social cancers that are wars and violence. Being also a scientist, he felt compelled to carefully inspect the validity of his beliefs. He was surprised to learn that in most countries more people die from suicide and many more from accidents than they do in wars. Overall, wars caused only 1.6 percent of all deaths, prompting him to conclude that "this is less than one might have guessed from the large amount of attention which quarrels attract." He felt compelled to go even further, adding with dismay that "those who enjoy wars can excuse their taste by saying that wars after all are much less deadly than diseases," a sobering conclusion for a Quaker who'd devoted such energy to preventing conflicts.

Despite his failure to predict the roots of violence, Richardson's work puts our historical drama in a new perspective. To start with, the massacre at Apátfalva's ford bucked conventional military wisdom. The crusader outposters and their purported allies—the cavalry that cruelly sliced them down—were unlikely foes by any accepted standard: The two sides were united by a common enemy, the Ottomans; they spoke the same language, Hungarian; and respected an overarching authority, the king, all circumstances that theorists claim lessen the likelihood of violent conflict. Richardson, of course, showed us that the theorists were wrong and that their predictions were only a myth. So the Apátfalva conflict is a textbook example of Richardson's legacy, violating all perceived laws of war and peace.

But if Richardson is right and the timing of wars and conflicts cannot be predicted, how did István Telegdi, the king's trusted advisor, dare to formulate his prophecy? It's quite possible that Telegdi ventured into the future with the audacity of the ignorant, not knowing that there is no basis for his foresight. But does Richardson's conclusion make our prophet inevitably naive and wrong?

The great thing about past conflicts is that there is little ambiguity about their future: We can simply open up the history books and check what happened next. Therefore, unlike his contemporaries, we can ascertain the accuracy of Telegdi's prophecy by scrutinizing the historical record. So let us do just that and check in on György Székely's reaction to the surprise attack on his outpost. Random as the massacre may have seemed, it was bound to have major consequences. It was the first time in this Crusade that weapons were drawn for real. It was the first test of György Székely's leadership, probing also the might of his army.

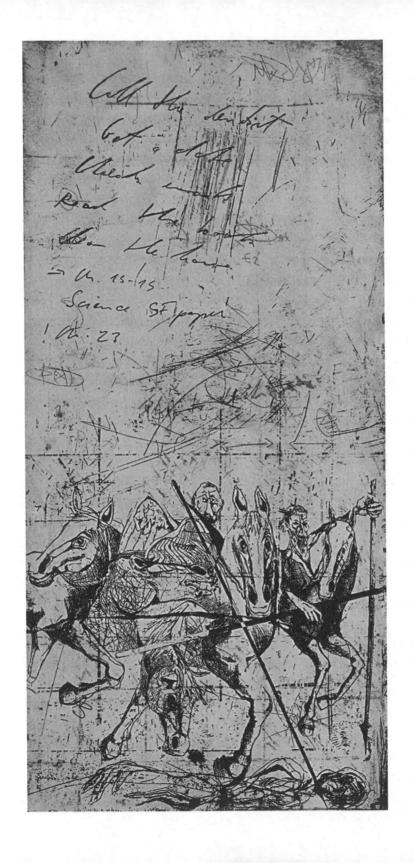

12

The Nagylak Battle

Nagylak, May 24, 1514, the day after the massacre

 blood-soaked riverbank scattered with mutilated bodies—that's all Báthory's army left behind at the Apátfalva ferry. The victorious cavalry and their mercenaries did not go far, however. They stopped for the night at the nearby Nagylak Castle, where, according to their contemporary chronicler, they "set up camp, played the zither and the shepherd's pipe, and sang and rejoiced." Did they know that it was the crusaders' outpost they had just massacred? It's hard to believe, but the historical consensus is that they were probably clueless.

If György Székely had continued to march his troops in the direction that he had been following over the first two weeks of the campaign, by May 24 his army would have been three to five days northeast of Apátfalva, approaching Transylvania. Following the cardinal's letter, remember, he changed direction, turning his army toward Belgrade. It is not clear if Báthory, busy gathering his forces in southeast Hungary, knew that the plans had changed. So who exactly did he think he'd just slaughtered? He could not possibly have mistaken the peasants for Ottomans.

Opposed to the Crusades from the beginning, the cantankerous cleric who'd shut the door in Székely's face, Bishop Csáky, failed to proclaim the papal bull in his episcopate, effectively forbidding recruitment. Word of the Crusade got out, nevertheless, and many serfs chose to join the campaign regardless. Perhaps, then, the bishop thought the Apátfalva camp an illegal gathering of peasants wishing to sign up as crusaders, against his will. And with the help of Báthory, perhaps he decided to teach the serfs a lesson that would resound across the land.

If Báthory, an experienced soldier, had known that György Székely's vast army was breathing down his neck, a mere day's march behind the bloody scene at the ford, it seems highly unlikely that he would have attacked the outpost. Even less probable is it that he would have allowed his men to *sing and rejoice* all night, failing to take basic military precautions in the dark of night. Evidently he was supremely confident as a warrior and tactician, because "do nothing" is apparently exactly what he did.

A few crusaders, including the leader of the outpost, managed to escape the bloodbath at Apátfalva, informing György Székely of the attack. Without hesitation György put his men to marching toward Báthory's camp. By the time he arrived, however, the local peasants had taken it upon themselves to avenge the massacre. During the cover of night they had filled the moat surrounding the castle's wooden walls with cords of dry tree branches and wood shingles pulled from the surrounding houses. Soundly asleep after their late-night celebratory binge, the nobility wasn't aware of danger until the loud cracking of the fire roused them. But by then it was too late to defend their retreat. The flames quickly engulfed the castle, and smoke exhaled from the walls, forcing the inhabitants to flee the fortified compound. Not suspecting foul play, they streamed out through the main gate, straight into the arms of the enraged villagers.

In the confusion, Báthory escaped through a side exit. Barefoot, wearing only his mantle, he hurried to his men, camped on the field next to Nagylak Castle. But by the time he had positioned his troops for battle, dawn broke, and he saw that the angry group of peasants had been joined by thirty thousand crusaders, ready to face their first test of battle.

The ensuing clash was no longer a lopsided matchup between a small outpost, taken unawares, and a group of battle-tested, well-armed knights. Two of the country's strongest forces were facing each other. What the peasant army lacked in experience they compensated for in numbers and resolve, formidably matched to the nobility's well-oiled war machinery. The former mercenary of the Ends, György Székely, versus the seasoned István Báthory, the cardinal's rumored second choice to lead the Crusades. If this was all just a big misunderstanding, now was the moment to clear things up, avoiding another bloodbath. The two forces were supposed to be on the same side; they did not have to clash.

They did, nevertheless.

When the two armies engaged, the nobility's cavalry had an advantage with their strong horses and long lances and quickly killed the first rows of peasants. But behind these fallen there were tens of thousands more serfs. Lost in a sea of crusaders, the knights were forced to fight the peasants one-on-one, turning the battle into the quagmire Báthory had hoped to avoid.

Despite Székely's having the upper hand, hours passed without a clear outcome. When finally the crusaders' victory seemed imminent, Báthory's guns came to life and, with their frightening thunder, turned back the tide. Though the artillery caused more noise than damage, the peasants, who had never before faced such beasts, were at a loss.

As the crusaders wavered, the difference between experience and enthusiasm began to show. The battle-tested mercenaries pressed their advantage, and the peasants were soon retreating in a disorderly fashion, many simply running for the cover of the nearby forest.

"Runs the peasant, and the cavalry right behind, chasing and cutting them," recorded the chronicler Taurinus of the sudden turn of fortune on the battlefield.

Following Hungarian tradition, Bishop Csáky's cavalry had been kept from engaging in the battle, serving as reinforcements should the nobility be overpowered. But blinded by the imminent victory and greedy for their share of credit, they now joined the chase without waiting for Báthory's order.

Then, just as all seemed lost, György Székely emerged unexpectedly from the middle of his retreating forces. Taurinus paints the scene once again:

> The brave Székely suddenly appears in the midst of the frightened,
> confused crowd, encouraging, nudging his men.
> He rewards some and persuades others
> not to tremble or fear death.

As he "gather[ed] his forces and [led] them in battle," the tide turned once again. Surprised by the peasants' newfound strength, just as abruptly as they had come Csáky's forces abandoned the fray, galloping toward the bishop's home in the nearby Csanád. According to our chronicler, their cowardice sealed their fate:

> Finally the noble forces were broken and injured
> by legions of commoners.
> Flying arrows consumed them
> Ambushed by spears and troubled by the reed
> Where the hiding peasants stabbed them to death.

Báthory took a blow to the temple and lost consciousness after falling off his horse. Later that night, when he awoke from his coma, he found himself in a pool of blood flowing from his nose. Under the cover of dark he managed to slip away, leaving behind a battlefield scattered with corpses. He took refuge in the swamps of the Maros River, and after the search for survivors subsided he escaped on a loose horse.

Bishop Csáky was not so lucky. According to Taurinus, he

> plunged in a deep ditch,
> trembling, poor soul, he hid there all night.
> They found him and took him into the custody of the cruel Székely.

The battle at Nagylak was a critical test for György Székely and his inexperienced crusaders, demonstrating much to everybody's surprise that they were a match to the nobility's war machine. Their victory showed

György's ability to turn the tide just when everything appeared to have been lost. Even the nobility's troubadour, Taurinus, former secretary to Cardinal Bakócz, was forced to acknowledge the charisma of the captain, calling him the "brave Székely."

Yet this victory was not without irony, for it was not a win against the Ottomans but serfs and peasants overpowering their own lords.

"To anyone brought up, as the author was, in the belief that Christians worship the Prince of Peace, it must be startling to notice the predominant occurrence of wars involving Christians, and especially of those in which Christians fought each other," wrote Lewis Fry Richardson in his *Statistics of Deadly Quarrels*. Indeed, he was unnerved to find that of the 128 conflicts between warring parties of the same religion, in 119 Christians fought Christians. Furthermore, of the 134 wars between differing religions, in 105 one of the parties was Christian. So perhaps it wasn't so odd at all, but a law of history, that the crusaders, their desire to battle the Muslims notwithstanding, should launch their campaign by massacring fellow Christians.

Nevertheless, the crusaders' victory was a remarkable feat: Never before in the history of Europe had an army of peasants defeated the country's official army in open battle.

Starting the Crusades in April, too early in the calendar year for a military campaign, was not the cardinal's only misstep. His other error in judgment was delegating the recruiting of the troops to the Franciscans without supervising their message. Sworn to poverty, the Franciscan priests had always shown sympathy toward the poor and the serfs. In fact, in an attempt to curb this natural alliance, monks were even excommunicated for writing and speaking out against the aristocracy's injustices. But with the cardinal's Crusade, in 1514 the Franciscans finally saw the opportunity to preach openly the idea they could previously only whisper in the privacy of their cloisters: It is against God's will that the landowners exploit the serfs. With the encouragement of their priests, the daily Masses in the camps turned into hotbeds of dissent, peasants and serfs angrily voicing their grievances.

The monks also made sure that everybody understood the papal bull: Those hindering the Crusades—father who holds his son back from the battle, son who holds his father back—were the enemies of God. So in the eyes of the crusaders, the bishop, Báthory, and their men were now the infidels. Their unprovoked attack had been contrary to the will of the cardinal, the pope, and God himself.

The king gave land and villages to lords and knights so that they could bring their own weapons, armor, horses, and men to bear whenever the king called them into battle. But if the wealth and tax-free status of these men was based on their military service, why, then, were the *peasants* now being asked to march to Constantinople? Why wasn't the nobility fulfilling its historical role and confronting the Ottomans alone? As far as the peasants were concerned, the battle lines were becoming increasingly clear.

But György Székely was not a peasant. As a Szekler, he was nobility by birth, free of the obligation of taxes, with his own coat of arms and his share of land in Transylvania. As a soldier he had served the crown for years. His nobility had recently been reaffirmed by the king, and he was now the proud squire of a nearby village. He was a loyal ally to the prevailing order, holding the trust of the cardinal and the king.

His actions were legitimate and defensible within the scope of his mission. He had been asked to train the peasants, and he had turned them into formidable fighters, as they had just now demonstrated. He had been ordered to take his army to Constantinople, and now they were halfway to the border. He did not attack the nobility—it was Báthory's army that had massacred his men. By fighting back, György Székely did what he had signed up to do, clearing his path to Belgrade. It was Báthory and Csáky who had defied the cardinal and king, and György saw their defiance for what it was—a political move to oppose the Crusades and his superiors' will.

While György Székely's actions were justifiable, the options available to him had been risky and thus debatable. On the one hand, the bishop and other aristocrats captured by his men were guilty of treason. If he had left their fate in the hands of his peasants, the outcome would not be in doubt—the angry mob would have promptly executed them. But it is one thing to kill in battle, in self-defense, and it is something altogether different to let the mob judge the cream of the aristocracy. Even if the king and the cardinal should forgive his aggression, this act would never be forgotten by the extended families of Csáky, Ravaszdy, Dóczy, Torpay, Orosz, and Tornallay, the prominent nobility now fearing their fate in captivity. But they were responsible for the deaths of thousands of crusaders, so releasing them was not an option, either, as it could mean a revolt in his camp.

Until now Székely's priorities had been clear-cut and easy: Recruit, train, and march the men to war. With the long years spent at the Ends and the countless battles he had fought, he felt quite confident of his ability to fulfill his mission. He was sickeningly unprepared, however, for the dilemma he now faced.

While the decision was his alone, he was not alone. There were those in his camp who would be satisfied only with the heads of the prisoners. And there was his brother, Gergely, at his side, cautious as always, pleading for his brother to wait and see. And there he was, György, in the middle, sensing that his decision would be momentous. Yet in order to choose the right path, he needed to set his priorities right. And as we will see next, once priorities come into play, randomness is out, and bursts take their place.

13

The Origin of Bursts

The manager of the first billion-dollar conglomerate in history, Charles Michael Schwab got his start as a stake driver for one dollar a day in Andrew Carnegie's steel mill. Yet the "master hustler," as Thomas Edison once referred to him, became an international celebrity only after he broke the bank at Monte Carlo in 1902. His two-hundred-thousand-dollar "gift" to the mistress of Grand Duke Alexis Aleksandrovich made him the steel supplier of the Trans-Siberian Railroad, where he earned enormous profits from the sixty-five thousand tons of rails he shipped to Russia. While Lewis Richardson was busy carrying the wounded off the French battlefield, Schwab was engrossed in circumventing American neutrality laws by smuggling the British just about anything they were willing to pay for, including twenty submarines.

When not absorbed in risky deals, Schwab had another obsession: efficiency. After his arrival at Bethlehem Steel in 1903, a blast-furnace superintendent called him crazy for setting unreasonable production targets. Most managers would have fired the rebellious employee, but

Schwab did not. Instead he publicly challenged the man, vowing that he would pay off his mortgage if the superintendent got the furnaces operating at the efficiency he, Schwab, desired. A few months later the superintendent had a mortgage-free house, and Schwab's furnaces were running at his own pace.

Given his addiction to efficacy, Schwab had a hard time refusing a proposal made him at a party by Ivy Lee, his long time publicist.

"I can increase your people's efficiency—and your sales—if you will allow me to spend fifteen minutes with each of your executives."

"How much will it cost me?" asked the shrewd businessman immediately.

"Nothing, unless it works," replied Lee, adding, "After three months, you can send me a check for whatever you feel it's worth to you."

Schwab took the deal and three months later mailed a thirty-five-thousand-dollar check, worth more than seven hundred thousand today, paid to the order of Ivy Lee.

What could Ivy Lee possibly have done in fifteen minutes to earn such a handsome payout? He approached each manager in Schwab's company, asking him the same thing:

"I want you to promise me that for the next ninety days, before leaving your office at the end of the day, you will make a list of the six most important things you have to do the next day and number them in their order of importance."

"That's it?" some asked incredulously.

"That's it," responded Lee, adding, "Scratch off each item after finishing it, and go on to the next one on your list. If something doesn't get done, put it on the following day's list."

I have been obsessed for years with priority lists. Of late I write them on the cardboard that the dry cleaners fold around my shirts—it is hard, sturdy, and survives for a full week. Absent my favorite cardboard, I often scribble priorities on anything that falls in my hands. Backs of envelopes. Post-it notes. On the sides of research papers or magazines. One of the little pleasures of each day is the almost ceremonial act of crossing off the completed tasks.

Recently I discovered that my obsession is by no means unique. Indeed, Ivy Lee's legacy is at the core of most time-management books and courses. Take, for example, Eugene Griessman's *Time Tactics of Very Successful People,* which has a chapter enjoining you to "Create a To-Do List That Works, and Work Your List." Then there's *Time Management* by Marshall J. Cook, who put his message, "Setting Priorities," right on the cover. I would never have guessed, however, that my fixation with priority lists would provide the key to another obsession that took over my life in 2004: finding the origin of the mystifying bursts that permeate human activity.

❧

Striking similarities between events of rather different natures are often simply explained. I learned this in 1999, when I was struck by the omnipresence of hubs in several real networks. They were the Kevin Bacons of Hollywood, stars who act with an extraordinary number of other actors, or Web pages like Google.com or Amazon.com with millions of links to other documents on the Web. These highly connected nodes were not there by accident, I soon learned, an insight that would help us discover the laws that govern the evolution of many real networks.

History was now repeating itself: By mid-2004 my lab had observed a series of puzzling similarities between events of quite different natures, seeing bursts and power laws each time we monitored human behavior. So now, the unexplained likenesses between our e-mail, Web-browsing, or printing patterns demanded an explanation. The rest of the summer, while visiting my family in Transylvania, I kept telling myself that there had to be a simple explanation to all of this! But my relentless probing produced nothing.

Scientists are often portrayed in books and Hollywood movies as furiously filling blackboards with complex formulas in an effort to find the answer to the Next Big Question. The truth is that often, not knowing where to start, we do nothing. This pretty much summarized my efforts to make sense of the ubiquitous bursts. I knew that this could not be a series of coincidences. But what's behind it? Is it math, physics, medicine, psychology, or social science? Bursts were pervasive, and that is exactly what made them perplexing, offering few clues as to where I could start deciphering them.

⁂

On the evening of July 2, 2004, I went to bed early, knowing that I would be getting up before dawn the next morning. I would take a cab to the Bucharest airport, five hours from Csíkszereda, my hometown in Székelyland, Transylvania, en route to a conference in Bangalore. Yet, the excitement of my first trip to India kept me from sleep. In that precarious twilight zone, not yet asleep but not really alert, either, my mind raced through the numerous tasks ahead of me. The talk in Bangalore. Passports. Bursts. Something to read on the long flight. Cash. Bursts again—where do they come from? Malaria medicine.

Jumping back and forth between the different items on my priority list and the research problem that absorbed me, I was suddenly struck with an idea, a simple explanation for the omnipresent bursts. And with that, my brain ended the frantic race, making me realize that it was not the details of the trip that had been keeping me awake but the mystery of the power laws in human activity. Having found an answer to the question that had haunted me for months now, with the high hiss of a tire punctured by a rusty nail, the pressure in my brain vanished, and I finally fell asleep.

The next day, after my son got lost in *The Jungle Book* on the airplane, I had my first chance to return to the idea I had dreamed up the night before. It was a model that had to be tested on a computer. There was one problem, however: I had failed to install Fortran on my laptop, the ancient programming language I still use in my research. I had only Mathematica, a software package that I had only used previously for visualization and quick math, but which could be used for scientific programming as well. I had a choice to make: I could either wait until I returned to the United States a month later, performing the calculations in my office, or I could learn programming in Mathematica, quite literally on the fly.

There wasn't much of a question as to which route I would take. Restless to validate the spark I'd seen in the twilight zone, on the nine-hour flight from Frankfurt to Bangalore I turned the idea into a Mathematica program. When the computer finally delivered the results, I was in for quite a disappointment: a plain-old Poisson-like random activity pattern filled the laptop's small screen. There weren't any bursts at all.

It happens to all of us—just before falling asleep, the solution to the problem that troubled us for days or weeks comes to us. It must be an ingenious self-defense our brain has been equipped with, helping us rest even under the most stressful of circumstances. In the morning after such a revelation I am often left wondering, *What was the miracle solution that put me at ease again?* Most often I cannot remember it, and I doubt that it was more than a mirage. Other times, confronted by the bright light of the morning, the answer doesn't sound all that brilliant after all.

The model I dreamed up on July 2 was another of those houses of cards that fall apart once confronted with light. Thus, I put the problem aside and focused instead on the conference and the subsequent visit to the Nagarhole National Park; the search in the jungle for the elusive tiger; the food that was "not spicy at all," according to our waiters, yet carpet-bombed our taste buds; and our cheerful driver, Babu, who, after a weeklong trip, during which he skillfully navigated our Tata between thousands of bikes and rickshaws, holy cattle, and pedestrians balancing a variety of objects on their heads, shared with us the reason he loudly sounded the car's horn every ten seconds: "In India, if the car's brakes are broken, that is okay—you can drive," he said calmly. "If the horn is broken, you must not. Very dangerous."

❧

My twilight-zone idea had a simple premise: We always have a number of things to do. Some people use to-do lists to keep track of their responsibilities, while others are perfectly comfortable keeping them in their heads. No matter how you track your tasks, you always need to decide which one to execute next. The question is, how do we do that?

One possibility is to always focus on the task that arrived first on your list. Waitresses, pizza delivery boys, call center operators—just about everybody in the service industry practices this first-in-first-out strategy. Most of us would feel a deep sense of injustice if our bank, doctor, or supermarket gave priority to the customer who arrived *after* us. Yet Ivy Lee did not ask the managers to write down *all* things that needed attention. He asked them instead to list the "six *most important* things" that they had to do the next day in "their order of importance."

In other words, he instructed them to set priorities.

The idea that put my brain to rest on July 2, 2004, was deceptively simple: Burstiness may be rooted in the process of setting priorities. Consider, for example, Izabella, who has six tasks on her priority list. She selects the one with the highest priority and resolves it. At that point, she may remember another task and add that to her list. During the day she may repeat this process over and over, always focusing on the task of the highest priority first and replacing it with some other job once it is resolved. The question I want to answer is this: If one of the tasks on Izabella's list is to return your call, how long will you have to wait for your phone to ring?

If Izabella chooses the first-in-first-out protocol, you will have to wait until she performs all of the tasks that cropped up before you. At least you know that you have been treated fairly—all other items on her list waited roughly the same time. Should Izabella decide to pick the tasks in order of importance, however, fairness is suddenly history. If she assigns your message a high priority, your phone will ring shortly. If, however, Izabella decides that returning your call is not at the top of her list, you will have to wait until she resolves all tasks of greater urgency. As high-priority tasks could be added to her list at any time, you may well have to wait another day before hearing from her. Or a week. Or she may never call you back.

I was hoping that prioritizing—halting some tasks and green-lighting others—could somehow explain our burstiness. My laptop told me otherwise: Despite prioritizing, most tasks waited about the same amount of time on our lists, following the distribution that Poisson derived more than a century earlier.

❧

Poisson distribution. Poisson process. Poisson equation. Poisson kernel. Poisson regression. Poisson summation formula. Poisson's spot. Poisson's ratio. Poisson bracket. Euler-Poisson-Darboux equation. This is only a partial list, and yet it shows the degree to which Siméon-Denis Poisson's work has impacted just about all branches of science. But what is so impressive is not the volume of his contributions but rather their depth, raising a puzzling question: How did Poisson manage to work simultaneously on so many quite different problems and yet stay sufficiently focused to offer deep and lasting contributions?

Well, he had a secret: a notebook and a tiny habit.

Each time Poisson encountered a problem he thought fascinating, he would resist the temptation to savor it. He pulled out his notebook instead and made a note of it and promptly returned to the problem that had absorbed him before the interruption. Once he solved the problem at hand, he mulled over the list of problems scribbled in his notebook, then picking as his next challenge the one he found the most interesting.

Poisson's little secret was lifelong, careful prioritizing.

Which makes matters even more puzzling: Prioritizing was behind Poisson's phenomenal success and earned an exorbitant consulting fee for Ivy Lee, yet my Mathematica algorithm during my flight to India was telling me that prioritizing has no impact on the execution time of the tasks. Essentially, if Schwab's managers or Poisson had rolled dice to decide their next task, overall each item on their lists would have waited approximately the same time. This, to put it mildly, did not make any sense.

~e

After two weeks in India, my mind was still racing around the burstiness problem. Convinced that I was on the right track, in the quiet of my mother's home in Csíkszereda, I decided to perform a series of careful checks. As I inspected the algorithm over and over, it looked as if it was generating bursts. Yet, paradoxically, the laptop continued telling me that the time it took for each task to be executed followed not the expected power law but a Poisson distribution instead. Again, this did not make any sense.

It took a few more hours of digging to realize that while the model was running just fine, the part of the algorithm that displayed the results had a tiny error in it, rooted in my unfamiliarity with the programming language I had used. After I fixed the bug, to my pleasant surprise the much-desired power law, the mathematical signature of bursts, appeared on my screen. In sum, a tiny programming mistake kept me from my eureka moment for several weeks.

In the end, the model consisted of a list of tasks, each randomly assigned a priority. Then I repeated the following steps over and over:

a) I selected the highest-priority task and removed it from the list, mimicking the real habit I have when I execute a task.
b) I replaced the executed task with a new one, randomly assigning it a priority, mimicking the fact that I do not know the importance of the next task that lands on my list.

The question I asked was, How long will a task stay on my list before it is executed?

As high-priority tasks are promptly resolved, the list quickly fills with low-priority tasks. This means that new tasks often supersede the many low-priority tasks stuck at the bottom of the list and so are executed immediately. Therefore, tasks with low priority are in for a long wait. After I measured how long each task waited on the list before being executed, I found the power law we had observed earlier in each of the e-mail, library, and Web-browsing data sets. So the model's message was simple: If we set priorities, our response time becomes rather uneven, which means that most tasks are promptly executed and a few will have to wait forever.

Despite its ability to spur productivity, prioritizing is not without ramifications. Its most important side effect is queueing, a phenomenon poised to emerge every time there is a shortage of something. A lack of tables in the restaurant; an insufficient number of agents answering calls at a customer-service number; a paucity of seats in a movie theater. The many little delays do add up—according to some estimates the average American spends two to three years of his life waiting for some scarce resource.

Our tasks and responsibilities are poised to queue thanks to a shortage of time. If we could simultaneously work on an arbitrary number of tasks, no one would need a priority list. Time is our most valuable nonrenewable resource, and if we want to treat it with respect, we need to set priorities. Once we do that, power laws and burstiness become unavoidable.

The effectiveness of prioritizing is only partly rooted in our moving the important tasks to the top of our to-do lists. The rare events, the long delays, are just as critical to the process. These are the outliers, and paradoxically they are not the tasks we resolve quickly but those that wait almost forever on our list. A *New Yorker* cartoon captured the sentiment: A businessman calmly says into his telephone, "No, Thursday's out. How about never—is never good for you?"

If you want to get things done, you have to be ready to say no occasionally. Indeed, priority lists work only if we are willing to make hard choices, separating the inconsequential tasks from those that truly matter. Done correctly, these lists help us turn the distracting cacophony of chores into the outliers that queue forever, helping us to focus our attention on the tasks that are of true significance.

✧

Back to Cardinal Bakócz. His aspiration to reconquer Constantinople was somewhat of a strange desire for a seventy-two-year-old prelate. Short of the papacy, he had everything he'd ever wanted—wealth, power, and influence, both in Buda and in Rome. So why accept a responsibility that would so profoundly disrupt the status quo he so enjoyed? Upon closer inspection, it turns out that in the cardinal's case we are dealing with an accidental priority—a role he had never sought.

Three years before the conclave, several cardinals rebelled against Pope Julius II and formed the Council of Pisa to elect a new pontiff. As they tried to win Cardinal Bakócz over to their cause, he toyed with their hopes until Pope Julius II dangled the papacy before him in exchange for his alliance with Rome. So the cardinal snubbed the rebellion, only to be double-crossed later by the ailing pope. Indeed, it was Julius II's last wish that "anyone but the Hungarian" be elected as the next pontiff that robbed Bakócz of the papacy.

Given the recent dissent, the new pope Leo X could not feel secure on his throne—particularly now, when his most powerful rival, the influential Cardinal Bakócz, had settled in Rome. So he invented a brilliant solution: Send his rival away, as far as the distant Constantinople.

Bakócz, however, was no fool. He recognized the pope's scheming and resolutely showed no inclination to leave Rome. So the pope had to sweeten the deal, offering Bakócz three additional episcopates, each a considerable source of wealth. The prelate remained unmoved.

On October 24, 1513, in a Vatican ceremony, the pope handed the cardinal the gold Crucifix of the Legate, marking the official start of the Crusades. Subsequently, in a splendid proceeding, the other cardinals walked Bakócz to the gates of Rome, a traditional show of support for an emissary departing on an important mission. Yet come November, the cardinal was still in Rome.

Finally the pope agreed that Bakócz could name a replacement for the helm of the Crusades as soon as he could find a capable leader. Assured that he was welcome to return to the Vatican as soon as the campaign was launched, Bakócz took the crucifix to Buda.

The critical fact is that at no point in time were the Crusades a priority, either for the cardinal or for the pope. Despite much pious sloganeering, this wasn't a struggle between Islam and Christendom, not a prelude to the "clash of civilizations." The pope wanted to see his rival gone, and Bakócz hoped to amass even greater wealth, power, and respect before the next conclave. In the end, the two powerful men's conflicting priorities saddled the full responsibility for the Crusades on the shoulders of a virtually unknown fighter from the Ends, György Székely.

György was now at a crossroads, however, confronted with the most important decision of the campaign: how to proceed following his victory over the nobility. Could he rise to the challenge presented him by the competition between the cardinal, the pope, and the nobility? Little did he know, but it would be the cardinal himself who would push Székely down a path predicted by no one but Telegdi. Indeed, while György was avenging his slain outposters at Nagylak, the cardinal, in an attempt to quell tempers, was composing a forceful letter. He was not aware of the tornados tiny butterflies can cause, so he had no way of knowing that he was about to trigger a revolution.

14

Accidents Don't Happen to Crucifixes

Nagylak, circa May 27, 1514, a few days after the massacre

ere are two crucifixes," shouted György Székely, struggling to make himself heard by the sea of men gathered before him.

"Those who disavow their faith and want to leave this camp, turn to the crucifix on the left," he said, pointing his sword to a crucifix tied to a tall pole on his far left. Then he aimed the sword to the right, to the opposite end of the vast grassland, adding, "The champions of the cross should join me under the crucifix on the right."

The peasants struggled to fully comprehend their captain's message. They knew that it had to do with the two letters he had just received, one from the cardinal and the other from the king himself, both in Latin, incomprehensible to all but the monks and the priests in the camp. One sentence, as translated by the monks, was crystal clear, however: "Until some later, more convenient time, the campaign is suspended," the cardinal had written. Yet it was hard to believe that he could have really meant it. Did this mean that the sacrifices they each accepted when they joined this holy war, defying their landlords and leaving behind their families, was worth nothing any longer? Did it imply that the death of thousands at Apátfalva and their victory at Nagylak, had all been for naught?

Yet, the facts were clear: The king and the cardinal had canceled the Crusades. And so now György Székely offered his men two alternatives, both of them devastating. Those who chose to gather under the left crucifix were free to leave and go home if so they wish. But making such a choice wasn't as easy as it sounds. While the king's letter offered protection from prosecution to those who left the camps, the crown's thick seal had little weight back home, where their landlord was their hangman and judge.

The other option, choosing the crucifix on the right and following their captain into his fight, was even worse. Carry on? If they were not to face the Ottomans, who would they fight? György Székely was just getting to that question:

"There is no bigger sin than men cruelly abusing their power to keep their own nation in servitude," he said, and he did not need to explain who the sinners were. If some had been confused when they joined the camps, the brothers and priests had enlightened them during the campaign.

But it was one thing to share their grievances with each other and another entirely to hear them from their captain, whose authority came directly from the cardinal and the pope.

"The sun shines equally on people and cattle alike, but you are lucky if the nobility lets you enjoy its benefits," continued György, pouring salt on the wound.

Many serfs and peasants had joined the Crusade with the vague hope that if they excelled on the battlefield their hardship back home might also come to an end. After all, the nobility's right to land was based on their military might. If now they, serfs, peasants, and outlaws, were to take on the Ottomans, would they not also be entitled to the same treatment and rights? But now, as the Crusade came to a sudden end, their last hope for a better life had vanished into the hot summer air.

The Székely knew exactly how they felt, so he made a conscious appeal to their anger and pain: "Go ahead, attack your enemies as long as fear benumbed them. Kill and chase away each of them!" he shouted with fervor. "Teach those senseless savages a lesson, to live the same way as their serfs and fellow citizens, to stop ruling with distorted pride and arrogance! Don't let the opportunity slip! Fight for your freedom!"

Many peasants had joined the Crusade out of religious zeal, believing it made them privy to the salvation the pope and the cardinal had promised. So the Székely now played on these religious sentiments—"Careful, don't anger God by missing this opportunity"—and offered himself as the leader of the next battle. "I promise to lead you, and with God's help, I will reconquer your freedom!"

When György Székely finished speaking, it was their turn to decide. Should they go left, choosing the road home and accepting everything as it had been before? Or turn right and join their captain in his fight against their oppressors? They watched the captain turn his horse and slowly approach the right cross, signaling his choice.

Volunteering to fight against the Ottomans with the blessing of the pope and the cardinal was one thing. Taking up arms against the aristocracy and accepting excommunication, as the cardinal promised those who refused his order to return home, was a completely different proposition. Their captain was now asking them to ignore both Constantinople and the cardinal and instead fight for their own rights.

Some were tired. Others were taken by a superstitious fear. Székely's vision was not what any of them had signed up for, so they started moving in opposing directions.

A few, who had already burned all the bridges there were to burn, followed their captain to the right. As they approached the crucifix, they watched with concern the thickening crowd on the left.

The Székely also viewed them with unease. He couldn't put up a meaningful fight with the handful of loyal followers that were heading his way.

He had been outvoted and abandoned. First by the cardinal, then by the king, and now by his own men. There would be no glory at the walls of Constantinople, no Te Deum for him in the church in the Buda Castle. He would be lucky to keep his head atop his neck at the end of this whole adventure.

Then, unexpectedly, the crucifix on the left fell off the pole.

An accident, perhaps. Too hastily tied in the heat of the events, maybe.

The monk guarding the post lifted it without delay and tied it back to the end of the pole. It took only a minute, a meaningless episode to most, and so the procession to the left continued undisturbed.

Then the crucifix fell again.

Once an accident. Twice a sign? *Isten nem akarja*. It's against God's will, said some.

Alert to its significance, the monk ceremoniously lifted the crucifix once again. He held it up like a sacred relic before tying it back to the pole.

A few in the thick lines heading to the left were at a loss, turning to their priests for guidance. *Isten nem akarja,* said others, and the word spread like wildfire, reverberating throughout the camp. Some merely stopped in their tracks. Others tried to escape from the left-heading columns. They broke the lines and turned the procession into a turbulent pandemonium.

In the midst of this chaos, the crucifix fell a third time, precipitating a throaty roar that morphed into a mad, rhythmic chant:

> *Isten nem akarja.*
> *Isten nem akarja.*
> *Isten nem akarja.*

Once an accident. Twice a sign. Third time can only be a miracle.

The monks fell on their knees. The peasants followed them.

Seizing the moment, the Székely cried, "Should God appear amid thunder and lightnings so that his horrible punishment for your incredulity will become an example to the world?"

Away from the cursed crucifix, the messenger of evil.

It's against God's will, they all chanted, and signed up for a war that none of them knew how to fight.

15

The Man Who Taught Himself
to Swim by Reading

A lbert Einstein, already a celebrated physicist but not yet the media star he was to become only a few months later, received a letter in the spring of 1919 from Theodor Kaluza, a virtually unknown colleague. Kaluza was still laboring to repeat the creative burst he'd enjoyed in 1908, when, as a student of David Hilbert and Hermann Minkowski, he'd written his first and only research paper. Now, a decade later and at the age of thirty-four, he was an unknown research drudge, still seated on the lowest rung of the academic ladder, barely supporting his wife and child on a practically nonexistent salary. When he finally finished his second paper, a rush of boldness prompted him to send it to Einstein, eliciting this encouraging reply on April 21, 1919: "The thought that electric fields are truncated . . . has often preoccupied me as well. But the idea of achieving this with a five-dimensional cylindrical world never occurred to me and may well be altogether new."

Today we teach physics students about five fundamental forces, three of which—gravitation, magnetism, and electricity—were already known to Kaluza and Einstein in 1919. For a long time these three forces appeared to have little to do with one another. In 1864, however, James Clark Maxwell showed that electricity and magnetism can be described by a single theory he called *electromagnetism*. His work inaugurated a dream that continues to drive the work of many physicists even today: Discover a Theory of Everything that captures within a single framework *all* forces of nature.

In 1919 the Holy Grail was a unified theory of gravity and electromagnetism. It was this problem to which Kaluza offered an unexpected solution when he showed that the two forces could be brought together if we assume that our world is not three- but five-dimensional. This hypothesis confounded intuition, naturally, as no human was ever able to perceive that enigmatic fifth dimension. But Kaluza had swum successfully on his very first try after reading a book on swimming, so he had no difficulty putting faith in theoretical knowledge this time, either, notwithstanding its counterintuitive nature.

Kaluza's letter to the scientific great was more than a mere courtesy—he was asking for Einstein to help him publish the manuscript. In those days, famous scientists like Einstein were the gatekeepers to the better scientific journals. If Einstein found the paper of interest, he could present it at the Berlin Academy's meeting, after which it could be published in the academy's proceedings. To Kaluza's joy, Einstein was willing.

Then, a week later, on April 28, Einstein wrote a second letter to Kaluza. The beginning was encouraging: "I read through your letter and find it very interesting. Nowhere do I see anything impossible yet."

What proceeded, however, was more subdued: "But I must admit that the arguments brought forward up to now do still seem to have far too little persuasive power."

After he probed Kaluza's arguments with a few technical questions and suggestions, Einstein added, "If you could show with precision established by our empirical knowledge that this is true, I would be as good as convinced of the correctness of your theory."

Einstein would open the academy's doors for Kaluza on one condition: "I could present a shortened version before the academy only when the above question of the geodesic lines is cleared up. You cannot hold this against me, for when I present the paper, I attach my name to it."

Imagine the feelings elicited in Theodor Kaluza after having received two letters in as many weeks from the man who was already considered the most influential physicist alive. Both letters were encouraging, and the very fact that Einstein had bothered to write twice indicated that he was genuinely taken by the unknown physicist's idea. But the letters also carried a large stick, which eventually prevented the paper's publication for years.

※

In 2005 I was invited by the Israeli Academy of Sciences to give a lecture in Jerusalem on the hundredth anniversary of Einstein's miraculous year. As the date for the trip approached, my anticipation grew for reasons not entirely related to the conference. Lately the priority model had precipitated intriguing follow-up questions. As you'll recall, we'd found that our e-mail correspondence was sprinkled with bursts of activity that soon tapered off to no activity, followed by frenzied e-mailing once again. And so, I'd begun to wonder, is the bursty pattern a by-product of the electronic age, or does it perhaps reveal some deeper truth about human activity? All the examples we had studied before—from e-mail to Web browsing—were somewhat connected to the computer, raising the logical question, did bursts precede e-mail?

I soon realized that the letter-based correspondence of famous intellectuals, carefully collected by their devoted disciples, might hold the answer to this question. An online search pointed me toward the Albert Einstein Archives, a project based at the Hebrew University of Jerusalem, whose mission was to catalog Einstein's full correspondence. As my e-mail to the archives remained unanswered, I tucked their address in my backpack, ready for my double mission: pay tribute to Einstein's annus mirabilis and find the Jewish National Library, the physical home of the Einstein Archives.

No need to go any farther, I was told at a reception some days later in Jerusalem. I had struck up a conversation with a group of prominent science historians also attending the conference, and when I mentioned my desire to study Einstein's correspondence, they pointed to the person standing just behind me. I was quickly introduced to Diana Kormos-

Buchwald, professor of history at the California Institute of Technology and the director of the Einstein Papers Project. She was able to confirm that the data I sought, though somewhat scattered, did exist and could be compiled. After we both returned to the States, she introduced me to Tilman Sauer, senior research associate in history at Caltech, who also worked on the Einstein Papers Project. A few messages and a couple of weeks later, Albert Einstein's complete correspondence, including his exchange with Theodor Kaluza, arrived by e-mail.

≈

In his response of May 1, 1919, Kaluza was quick to dispel Einstein's concerns, prompting this letter from Einstein on May 5:

> Dear Colleague,
>   I am very willing to present an excerpt of your paper before the Academy for the Sitzungsberichte. Also, I would like to advise you to publish the manuscript sent to me in a journal as well, for ex[ample] in the *Mathematische Zeitschrift* or in the *Annalen der Physik*. I shall be glad to submit it in your name whenever you wish and write a few words of recommendation for it.

What caused Einstein to change his mind so suddenly? His letter offers a hint: "I now believe that, from the point of view of realistic experiments, your theory has nothing to fear."

Kaluza could not have hoped for a better outcome. Einstein, well-known for his never-ending quest to confront all mathematical developments with reality, had accepted his conclusion that our world is five-dimensional. You and I may not have any real sense of this fifth dimension, but such lack of perception has never stopped a physicist, armed with the penetrating power of mathematical theories, in his quest to make sense of our universe. If the math says that the world needs a few extra dimensions, who are we to stand in its way?

≈

A rather prolific correspondent, Einstein left behind about 14,500 letters he had written and more than 16,000 he had received himself. This averages out to more than one letter written per day, weekends included, over the course of his adult life. Impressive though it was, it was not the volume of his correspondence that piqued my interest. In the spirit of the priority model, I wanted to find out how long Einstein waited before he responded to the letters he had received.

It was João Gama Oliveira, a bright Portuguese physics student visiting my research group on fellowship, who first studied the data we got from Caltech. His analysis showed that Einstein's response pattern was not too different from our e-mail patterns: He replied to most letters immediately—that is, within one or two days. Some letters, however, waited months, sometimes years, on his desk before he took the time to pen a response. Astonishingly, João's measurements indicated that the distribution of Einstein's response time followed a power law, similar to the response time we had observed earlier for e-mails.

João and I were first concerned that any long lapses in Einstein's correspondence were due to missing letters in the data set. Tilman Sauer at Caltech reassured us, however, that many of the long intervals corresponded to genuine delays. Take, for example, Einstein's reply on October 14, 1921, to Ralph de Laer Kronig: "In the course of eating myself through a mountain of correspondence I find your interesting letter from September of last year." And, indeed, the records showed that Kronig's letter had gathered dust on Einstein's desk for more than a year, unanswered.

But it wasn't only Einstein's correspondence that followed the pattern. From the Darwin Correspondence Project hosted by the University of Cambridge in England we obtained a full record of Charles Darwin's letters. Given that the meticulous Darwin kept copies of every letter he either wrote or received, his record was particularly accurate. Its analysis indicated that he, too, responded immediately to most letters and only delayed addressing a very few. Overall, Darwin's response time followed precisely the same power law as had Einstein's.

The fact that the records of two intellectuals of different generations (Einstein was born three years before Darwin's death) living in different countries follow the same law implied that we were not looking at the idiosyncrasies of a particular person but the basic pattern of preelectronic communication. It also meant that it is completely irrelevant whether our messages travel on the Internet at the speed of light or are carried slowly across the ocean by steam engine. What matters is that both back then and today we face a shortage of time. We are forced to set priorities—even the greats, Einstein and Darwin, are not exempt—from which delays, bursts, and power laws are bound to emerge.

Yet a peculiar difference remains between e-mails and letter-based correspondence: The exponent, the key parameter that characterizes any power law, was different for the two data sets.* This difference meant that there were fewer long delays in e-mail correspondence than in letter writing, not a particularly surprising finding, given the immediacy we often associate with electronic communications.

The truth, however, is that the difference could not be attributed to the delivery time. Decades of research have told us that the exponent characterizing a power law cannot take up arbitrary values but is uniquely linked to the mechanism behind the underlying communication pattern. That is, if a power law describes two phenomena but the exponents are different, then there must be some fundamental difference between the mechanisms governing the two systems. Therefore, the discrepancy insisted that a new model was needed if we hoped to account for Einstein's and Darwin's letter-writing patterns.

Heartened by Einstein's encouragement, Theodor Kaluza quickly made the requested changes and mailed back a shorter version of the paper, appropriate for presentation at the academy. His case looked really good now—he had received four letters in less than four weeks, indicating that the famous physicist had assigned him an unusually high priority. Yet, in a letter dated May 14, 1919, Einstein unexpectedly gave him the cold shoulder. "Highly esteemed Colleague," he wrote, "I have received your manuscript for the academy. Now, however, upon more careful reflec-

---

* In mathematical terms, in the power law $P(\tau) \sim \tau^{\delta}$, describing the probability $P(\tau)$ that a message waited τ days for a response, the exponent was $\delta=1$ for e-mails and $\delta=3/2$ for Einstein's and Darwin's correspondence.

tion about the consequences of your interpretation, I did hit upon another difficulty, which I have been unable to resolve until now."

In a four-point derivation, Einstein proceeded to detail his concerns, concluding, "Perhaps you will find a way out of this. In any case, I am waiting on the submission of your paper until we have come to some resolution about this point."

With that, he sent Kaluza back to the drawing board.

⁕

In our priority model, we assumed that as soon as the task of highest priority was resolved, a new task of random priority took its place. To derive an accurate model of Einstein's correspondence, we needed to modify the model, incorporating the peculiarities of letter-based communications. Indeed, with snail mail, each day a certain number of letters arrives by mail, joining the pile of letters already waiting for a reply. And so, whenever time had permitted, Einstein had chosen from the pile those letters he considered most important and replied to them, keeping the rest for another day. So a model of Einstein's correspondence has two simple ingredients:

a) With some probability, which we will call the letters' *arrival rate*, letters landed on Einstein's desk, increasing the length of his queue. He assigned some priority to each letter.
b) With some other probability, which we will call the *response rate*, he chose the highest-priority letter and responded to it.

If Einstein's response rate was faster than the letters' arrival rate, then his desk was mostly clear, as he had been able to reply to most letters as soon as they arrived. In this *subcritical regime* the model indicated that Einstein's response times followed an exponential distribution, devoid of long delays and clearly not in accordance with the observed power law.

If, however, Einstein responded at a slower pace than the rate at which the letters arrived, the pile on his desk towered higher with each passing day. Interestingly, it is only in this *supercritical regime* that the response times follow the power law observed earlier for Darwin and Einstein. Thus, burstiness was a sign that Einstein was overwhelmed, forced to ignore an increasing fraction of the letters he received.* But was he indeed swamped?

While only seven letters survive from his annus mirabilis, it is safe to assume that in 1905 the obscure patent clerk had no difficulty staying on top of his correspondence. Indeed, in those days only his family and a few close friends demanded his attention. One decade later, in 1915, he had already become a renowned physicist, which brought a number of responsibilities, prompting him to write a letter every other day or so. While only twelve of the letters he received in that period survive, they give no evidence that his correspondence placed extraordinary demands on him or was particularly delayed.

But if Einstein was able to keep up with his correspondence, as his timely replies to Kaluza in 1919 also suggest, then his response time should have followed an exponential distribution rather than the power law that we observed. That is, it should have been free of delays and bursts.

ೋ

Kaluza made one more attempt to persuade Einstein of the validity of his approach, even daring to point out an error in Einstein's arguments. May 29, 1919, saw Einstein's decisive reply:

---

* Why are the exponents different for the priority model introduced in Chapter 13 and the correspondence model explored here? Well, there is an important difference between the two models: the length of the priority list. In the priority model the number of tasks ahead of us was kept unchanged, as a new task entered the list only after one on the list was resolved. In the correspondence model, however, the length of the queue changes all the time, becoming longer each time a new letter arrives and shrinking whenever a letter is responded to. This may appear to be a tiny difference only, but the math showed that it is sufficient to change the exponent. But why not allow the number of tasks to change as new responsibilities arrive?

True, the number of tasks ahead of us surely changes with time. But are we aware of it? In 1967 George Miller published a landmark paper entitled *The Magic Number Seven,* in which he argued that our short-term memory is finite. We can easily remember seven numbers, but most of us fail to recall a twelve-digit string after one reading. We can remember a list of seven words, but fail to recall fifteen unrelated words. Miller's paper puts our priorities in a new perspective: We may have fifty tasks we should be paying attention to, but surely most of us cannot remember more than seven, plus or minus two. So our effective list of priorities does not fluctuate too much—our short-term memory has room for new tasks only when we have resolved the old ones. When it comes to the letter-based correspondence, Einstein did not have to stretch his short-term memory—the pile of letters on his desk served as an external memory, allowing him to keep an arbitrarily long queue.

Dear Colleague,

It is true that I made that blunder with the dS and ds in my introductory discussion. I see that you have also thought about this matter quite thoroughly. I have great respect for the beauty and boldness of your idea. You must understand, however, that due to my existing substantive reservations, I cannot support it *in the originally conceived manner.*

It is not easy for me to advise you on whether to publish the idea in its present form. I cannot even see beyond what we have reported to each other. Even so, publication of the findings up to this point is justifiable, especially if you point out any remaining problems. If you take this path and possibly have problems with the editors of the *Mathe[ematische] Zeitsch[rift]* (which I do not anticipate), I would be glad to put in a good word for you.

I am sending you my last paper, which, for lack of anything better, stops at the dualistic interpretation but is nevertheless of some interest, particularly with regard to the cosmolog[ical] problem.

With best regards, yours,

A. E.

Despite the polite tone, the rejection was clear, and we know of no more exchanges that year between Einstein and Kaluza. Nor in the following year. And not because Kaluza's paper was published. On the contrary, Einstein's reservations sent an unmistakable message to the young scientist: The fifth dimension was a blunder, either premature or a dead end not worth further attention. After a furious burst of communication had ricocheted between the two for a full month, a years-long silence followed.

∿

In 1915, four years before trading letters with Kaluza, Einstein had published yet another landmark paper, which had united relativity and gravitation. He had named his theory *general relativity,* and it was a thing of beauty, though a bit heavy on predictions and light on experimental confirmation. Finally, on September 22, 1919, four months after sending his last letter to Theodor Kaluza, Einstein received a cryptic telegram from the Dutch physicist Hendrik Antoon Lorentz. It read:

```
EDDINGTON FOUND STAR DISLOCATION AT SOLAR RIM
PROVISIONAL MAGNITUDE BETWEEN NINE TENTHS SECOND
AND DOUBLE-LORENTZ.
```

Einstein instantly made sense of the technical gobbledygook: The theory he'd posed back in 1915 had finally been confirmed with Arthur Stanley Eddington's observation that light is bent as it passes by the sun. Within days, Einstein's name was on the front page of newspapers and magazines all over the world, and the Einstein myth was born. He turned into a media superstar and immortal icon.

His sudden fame had drastic consequences for his correspondence. In 1919, he received 252 letters and wrote 239, his life still in its subcritical phase, allowing him to reply to most letters with little delay. The next year he wrote many more letters than in any previous year. To the flood of 519 he received, we have record of his having managed to respond to 331 of them, a pace, though formidable, insufficient to keeping on top of his vast correspondence. By 1920 Einstein had moved into the supercritical regime, and he never recovered. The peak came in 1953, two years before his death, when he received 832 letters and responded to 476 of them.

As Einstein's correspondence exploded, his scientific output shrank. He became overwhelmed, burdened by delays. And with that his response time turned bursty and began to follow a power law, just as our e-mails do today.

～

Despite his brief correspondence with Einstein, Kaluza's life improved little in the following years. He continued to work as a *privatdozent*, unable to find a position given his lack of publications. Then, two years after their last exchange, on October 14, 1921, he received this surprising postcard from Einstein:

> Highly Esteemed Dr. Kaluza,
> I have second thoughts about having you held back from
> publishing your idea about the unification of gravitation and
> electricity two years ago. Your approach certainly appears to
> have much more to offer than Weyl's. If you wish, I will present
> your paper to the academy.

And he did, on December 21, 1921, two and a half years after first learning of Kaluza's idea.

Why this sudden reversal? Had Einstein simply been distracted by his triumph, forgetting for years about Kaluza's extra dimension?

Forget he had not. The truth is that between 1919 and 1921 Einstein had been furiously searching to codify his version of the Theory of Everything, a unified theory of gravitation and electromagnetism. By September 1921 he had lost hope of ever succeeding along the lines he had been pursuing, a direction originally proposed by Hermann Weyl. Back at square one, Einstein remembered Kaluza's approach. And so in October 1921, in collaboration with Jacob Grommer, Einstein followed up on Kaluza's still-unpublished paper and came to an embarrassing conclusion: He could not continue blocking the publication of Kaluza's proposal while attempting to write his own paper inspired by it. So he finally released Kaluza's genie from its bottle.

By the time Kaluza's paper was finally published, it was too late for its author. Discouraged by Einstein's rejection, Kaluza had left physics and started anew in mathematics. But the professional switch eventually paid off eight years later when, in 1929, he was offered a mathematics professorship at Kiel University and in 1935 became professor at Göttingen, one of the most prestigious universities at the time.

Kaluza's and Einstein's brief encounter vividly illustrates that prioritizing is not without its consequences. Prioritizing meant Poisson's lifelong success in proving enduring theorems and Ivy Lee's windfall after he so successfully advised Schwab's managers at Bethlehem Steel. And prioritizing also led to the demise of a young physicist's career when his theories went ignored by the man who could have gotten them published. To be sure, Kaluza's multidimensional universe was eventually revived in the 1980s and became the foundation for string theory, whose proponents have no fear of five-, eleven-, or many more–dimensional spaces.

Kaluza did not live to see the renaissance of his work, however, passing away in 1954. Might he have turned into one of the greatest physicists if Einstein had allowed him to publish his breakthrough early on? We'll never know.

～

Letters and their consequences. They pertain even to our sixteenth-century events. For the fate of that Crusade also turned on two letters—one written by Cardinal Bakócz and the other by the king, both calling for an immediate end to the campaign. Given the string of events that preceded their writing, the directive they contained is by no means surprising. After all, the crusading army, assembled to liberate Constantinople, had just dealt a devastating blow to Báthory, effectively eliminating Hungary's southern defenses. And with that, the assemblage of serfs emerged as a terrible new force, more powerful than even the country's official military. What else could the king and the cardinal do but to put an end to the derailed campaign?

Until his receipt of those two letters, everything György Székely had done—including avenge his fallen outposters by decimating Báthory's forces—had arguably been part of his mandate to reconquer Constantinople. But on May 28, after the cross had fallen and his men vowed to follow him, György Székely refused to dismantle his army and executed Bishop Csáky and all the nobility in his custody. He was past the point of no return.

On May 27 he had been captain of the Crusades, in charge of a massive army on its way to confronting a most powerful adversary, the Ottoman Empire. By May 29 he was captain of the largest uprising that Hungary had ever seen.

In many ways this metamorphosis was not entirely his choice but had been forced on him. And if György Székely was angry about it, he had justification: As far as he was concerned, he had closely followed the spirit of his initial orders and the papal bull. Even if the cardinal couldn't condone Székely's actions, he should have at least understood they were in the interest of conquering Constantinople. Yet Székely was abandoned at the first sign of difficulty, scapegoated for everything that was to follow.

It is hard to pin down when and why György Székely changed from a Crusade captain into a freedom fighter. One thing is sure: On May 28, with the merciless execution of his aristocratic prisoners, the transformation was complete. He had freed himself of the cardinal and the king, taking his fate into his own hands.

But who was György Székely that he dared turn his sword on the very people who had created him? Why did his contemporaries never mention his family name but refer to him only by his ethnicity? What past deeds exiled him from Szekler Land and cast a dark shadow over his path to Buda, a past that everyone but the late Bishop Csáky had conveniently chosen to ignore? Was Csáky's execution a strategic move, baptizing György Székely into his new role, or merely the fulfillment of a personal grudge?

In light of his conversion from captain of an army that carries the blessing of the king, to leader of a social revolution condemned by the crown, perhaps Telegdi's warning to the king's court those months before take on new meaning: "The sword given to them to destroy the enemy, will they not turn it against us?" Indeed, was György Székely's sole role in history to fulfill Telegdi's prophecy? We can't rightfully settle on an answer yet, as we now must take a detour in order to better understand Székely. That is, we need to take an Einsteinian jump forward through time and space and visit the Transylvania of today, searching for the missing link, the reason why György Székely abandoned his home in the first place.

16

An Investigation

Nagyszeben, July 20, 2007

 no longer recall how many yellow stars I counted on the EU flag sagging in the tranquilizing July heat at the State Archives in Nagyszeben. I am sure, however, that there was no star representing the part of the world I was in—known as *Erdély* to the local Hungarians, *Siebenbürgen* to the German-speaking native Saxons, *Ardeal* to the Romanians, and, to the rest of the world, by its Latin name, *Transsilvania*.

I was in Nagyszeben to complete a journey, one that had started several years before almost by accident when my attention had been piqued by an obscure detail I noted among the 1514 events. To be sure, György Székely is known to everyone in Transylvania or Hungary, having become a symbol of the peasants' struggle against their feudal oppressors. Yet we always viewed him—together with the rest of the heroes that Communism sanctioned—with some degree of healthy suspicion. As I delved into the details of the 1514 events I was shocked to realize how different were the scenes I had learned during years of education in the

Communist system from those told by generations of less ideologically biased historians.

By this time in 2007 György Székely had squeezed himself into the narrative of this manuscript, prompting me to search for the document whose purported existence had started me on this journey. It was a letter written in 1507, seven years before György Székely's bursting onto the national scene, and it provided insights into his character and past that otherwise would have been lost to us.

Very little is known about his pre-Crusade life. In *Odeporicon*, the first written account of the Crusades, written one year later in 1515, Bartholinus Riccardus calls György Székely *Georgius Zechelius;* Taurinus's chronicle depicts him as a supernatural monster going by the name *Zeglius;* the contemporary Tubero calls him *Georgius Scytha;* and Szerémi in a chronicle written four decades after the battle refers to him as *Georgius Siculus* or *Zekel.* Each of these surnames refers to the *Szeklers,* the Hungarian-speaking tribe of the eastern Carpathians in Transylvania.

It was Istvánffy, in his chronicle written around 1605, who first whispered the family name, referring to him as *Dosa,* the Latin for Dózsa, a family from the village of Dálnok in the Szekler Land. The name stuck, and today that charmingly sleepy village in southeast Transylvania, squeezed into the valley of the minor Dálnok Creek, takes pride in being the birthplace of György Dózsa, the Székely. A small monument stands in a garden where the house he was born in once stood, and a massive statue hewn from rock depicting the knight dominates the modest village center. Other than his Dálnok origins, for centuries the only thing we knew of the pre-Crusade György Dózsa Székely concerned his duel at Belgrade. Then, in 1876, a rare find rewrote history.

On November 3, 1869, the Transylvanian chapter of the Hungarian Historical Society decided to collect and publish all written documents pertaining to the Szeklers' history. The job was somewhat capricious, given that the few times the Szeklers had expressed themselves in written form (rather than, say, with a battle-ax) they did so using a *rovás* script. This alphabet, devoid of vowels, is comprised of strange, sticklike characters that vaguely resemble the Middle Eastern syllable-writing system or the runic alphabet used in Western Europe prior to the adoption of the Latin letters. Inscriptions with the ancient *rovás* script can be seen even today in churches and on the richly carved, imposing gates wealthy Szeklers erect in front of their homes.

Unlike the *rovás* script used by the native Szeklers, most of the documents the Hungarian Historical Society was planning to collect were written in Latin by the Transylvanian aristocracy and clergy, the bureaucracy of the Middle Ages. It was during an effort to put these disparate documents in some kind of order that Károly Szabó, the project's founding editor, discovered the letter that put György Dózsa Székely's pre-Crusade life in a startlingly new perspective.

By the time I got a copy of the 1876 report on Szabó's discovery, the whereabouts of the original 1507 document were something of a mystery. Since 1876, Transylvania has changed hands several times, suffering more than a century of turbulence during which archives were burned and documents stolen, lost, and subjected to the ideological neglect of Communism. After I spent several months inquiring about the whereabouts of the document, on April 17, 2007, I received an enthusiastic message from Hédi Erdős, the librarian at the Institute of Advanced Studies in the Buda Castle. She told me that the original letter remained where Károly Szabó had discovered it 130 years earlier, in the Saxon National Archives in Nagyszeben, under the catalog number *Materia V, No. 67.*

When I arrived in Nagyszeben, I had little hope of actually seeing the document. As a youngster I had spent countless hours in another archive, that of the Mikó Castle in my hometown, Csíkszereda. The original castle, built around 1063, had been leveled by the Turks in 1661 and rebuilt in 1714 with thick walls and four bastions that stand today. For the past five decades the castle has housed the Museum of the Szekler Land, and in the 1980s my father had an apartment there as the director of the institution. Thanks to this somewhat unusual arrangement, I could freely roam through the museum's library and collections. This was out of the ordinary, to say the least—normally only historians who displayed sufficient loyalty to the prevailing Communist ideology were privy to the documents stored there.

While Communism had collapsed in Romania back in 1989, I knew from my yearly trips home that the reflexes of the old system continue to linger, making it almost impossible to access historical documents. My concerns were reinforced by Mark László-Herbert, a historian from the University of Toronto, who shared on his Web site his experiences with various Romanian archives. He wrote, "As a rule (well, not always applied), *foreigners* must get permission to conduct research from the Bucharest 'headquarters' before they arrive [at] a regional (county) section of the national archives." László-Herbert's caution did not worry me, since I was born in Transylvania and hence carry a Romanian passport. But a few lines later, my heart sank. "Romanian citizens residing abroad may have to follow the same procedures as foreign researchers," he continued. Given that I had been living in the United States for the past two decades, this meant me, of course. But I had no time to apply for a permit in Bucharest, and I suspected that, in going directly to Nagyszeben, I was setting myself up for a disappointment.

Sure enough, once I reached the Saxon National Archives, a policeman greeted me, demanding my *bulletin,* the Romanian national ID. *Here we go,* I thought, and nonchalantly asked if a passport would suffice. It did, and after entering my information into a long registry, he waved me up to the second floor, mentioning only in passing that the archives were closed that day.

What a bother, I thought. But after hours of driving in the late-July heat without air-conditioning, I was not yet ready to give up. Half a minute later the secretary confirmed that in anticipation of an inspection from Bucharest, the archives were, indeed, closed to visitors.

"Can I come back next week?" I asked.

"Not really," she responded, explaining that the inspection was to last a full week.

"The week after?" I remained hopeful.

"That'll be August," I was told, "and it's vacation."

Finally, eager to pass me off to somebody else who would have the authority to kick me out, she ushered me into the "closed" reading room, a small, sunny room with a high ceiling and six reading tables. Two office desks were strategically placed in two corners so that their occupants could keep an eye on the rare researcher who somehow gained entry to the well-guarded sanctuary. The room was empty except for a middle-aged lady in a neat white summer dress, carefully drawing calligraphic letters on a large poster laid flat on one of the desks.

I resurrected my most eloquent Romanian, rusty after a disuse of almost two decades, trying to impress on her that had I come all the way from America to see a single document. She nodded empathetically and confirmed that they were closed, but she did not send me away. She asked for my photo instead.

This was good news according to my source: "Upon your arrival, you will be issued a researcher card (have an ID/passport-size photograph on you)." Despite the clear forewarning, I had not brought a photo with me, so she directed me to the nearest photographer. Off I went into the beautiful center of Nagyszeben, the city that the native Saxons call *Hermannstadt* but that appears on international maps under its Romanian name, *Sibiu*.

The town was founded around 1150, when the Hungarian king Géza II invited five hundred families from the west of the Rhine to settle in Transylvania, cultivate the empty land, and protect the country's southeast border from Mongol and Tartar raids. A century later the thriving settlement aroused the Mongols' greed, and they promptly leveled the village. The surviving Saxons learned from the devastating experience and surrounded themselves with thick walls that no enemy since has been able to take by force.

Two world wars have somehow bypassed the charmed city, and Nagyszeben has emerged more or less unharmed from the centuries of turbulent Transylvanian history. Even the Communists, who under the pretext of progress had systematically decimated the country's heritage, had done little damage other than to leave the historical downtown in a state of desperate disrepair.

In 2007 Nagyszeben was one of the European Capitals of Culture, so it had been freshly restored to its medieval glory, thanks to the aid of European funds. I could have easily spent days there, wandering through the narrow roads around the majestic Gothic Evangelical Church that dated back to 1320, exploring the acclaimed art collection of Baron von Brukenthal, Transylvania's governor during the reign of Empress Maria

Theresa, or just taking in the warm July sun at one of the many outdoor cafés surrounding the fifteenth-century Grösser Ring, the town's central square. But there was no time for sightseeing. Following the directions I'd been given by the lady at the archives, I turned left on Heltauergasse, a thriving shopping street packed with galleries, quirky shops, and restaurants, and found the photographer.

Half an hour later, I handed the archivist the still-warm Polaroid shot, and she waved me to a seat, handing me a document request form. I scratched down the magic code *Materia V, No. 67,* given to me back in Budapest, asking her if she knew what these numbers meant. *Not really,* she shrugged, but took it, nevertheless, reminding me again of my Canadian guide: "Once you have found what you were looking for (if you are really, really lucky), you can fill out your slip and submit it. Then you will be asked to leave and return the following day."

Well, by now it was almost two P.M. on Friday, and there was no tomorrow, and no Monday or Tuesday, either, thanks to the impending inspection.

So there I was, an arm's reach from the EU flag, waiting for a Romanian bureaucrat to find a Latin letter written in 1507 by a Hungarian aristocrat about a Szekler fighter, preserved thanks to the caring guardianship of countless generations of Saxon archivists. I was anxious to learn if the document last seen in 1876 still existed, to see if *Materia V, No. 67,* meant anything over here.

Twenty minutes later, she reemerged, holding only a few loose sheets of paper, which looked awfully like the application I had filled out earlier. This was not good news, according to my source: "An archivist will pull the documents for you and bind them neatly in a binder (this is how—they say—they prevent the theft of documents; most often, however, this procedure visibly damages the document)."

So far my guide had been right about everything—the photo, the permit, the slip. So the lack of a folder was understandably disconcerting; it could mean only that my application had been rejected or that the scant information I had provided was insufficient to locate the document.

She came straight to my seat and, with a flick of her wrist, dropped onto my desk a small piece of brownish paper previously hidden by the sheets. Then, without a word, she resumed working on the large poster.

For a second I just sat there, staring at the thick paper folded in four. Touched by the unexpected kindness of the Romanian archivist, my hope was suddenly restored in the future of the four nations who had been cohabiting in Transylvania for centuries.

Could I touch it? Didn't I need some special gloves? Would the old document crumble in my hands?

Encouraged by the fact that nobody seemed to care or even be paying attention to me any longer, I carefully pulled the letter closer to me, trying to decipher the handwriting next to the dark spot of a long-gone wax seal.

"*Prudentibus et Circumspectis Magistro ciuium Judicibus Juratis ceterisque ciuibius at consulibus Ciuitatis Cibiniensis dominis at amicis honorandis,*" it said, and I wished that I had paid greater attention to my eighth-grade Latin teacher. One thing was encouraging, however: The mysterious code V and the number 67 were scribbled with black ink on the letter's back, offering hope that this just might be the link in the chain I was there for.

After convincing myself that the document was not about to crumble into dust, I carefully unfolded the five-by-seven-inch letter to find a page densely filled with handwritten text embellished with elegant arches and curves, conspiring to interrupt the mysterious monotony of the medieval Latin. These acrobatic undulations, together with the lines that started approximately one inch from the page's left margin and ended cleanly about a half-inch from the right, suggested that the letter's author had paid just as much attention to calligraphy as to his ideas.

I searched for a date on the letter without much success. Only later, with a transcript in my hand, did I learn that there was one, but not in the format familiar to us today. Instead of month/day/year, the last line before the signature read, ". . . on the first Monday after the feast of the Blessed Virgin and martyr Margaret in the seventh year"—meaning July 19, 1507.

In hindsight, this was an amazing coincidence: That day, as I sat in the archives poring over the aged text, the calendar showed July 20,

2007—that is, exactly five hundred years and one day after its author had taken pen to paper. By some strange providence I was in Nagyszeben precisely half a millennium after the mayor of Nagyszeben, to whom the letter was addressed, had broken the wax seal securing its contents.

At this point, as I was unable to discern the secrets of the mysterious Latin calligraphy, the only thing confirming that I held in my hands the right document was its signature, familiar and unmistakable:

17

Trailing the Albatross

In 2004, on his flight from Germany to Montreal for the March meeting of the American Physical Society, Dirk Brockmann couldn't have guessed that his professional life was about to change fundamentally. Even if he'd had an inkling of the significance of the journey he was taking, he couldn't possibly have foreseen that the inspiration for his career-changing discovery would come not from an insightful presentation at the conference he was about to attend but over a beer in the snowy setting of a Vermont forest, a few hundred miles south of Montreal. In any case, in the end Dirk felt compelled to thank his friend Dennis Derryberry for having introduced him to the currency-tracking Georgers. And so, in the acknowledgments of the paper he was about to publish in *Nature,* Dirk Brockmann graciously mentioned his friend's role in the discovery. Imagine his surprise when Dennis's title— cabinetmaker—was deleted from the text by the prestigious journal's copy editor.

"I was unamused about it, as it was my intention to have it in there," recalls Dirk, convinced that Dennis's profession was pertinent precisely because he was not a scientist. "So I inquired at *Nature* what the deal was. They said it does not comply with house style."

There was nothing unusual about Dirk's acknowledgment. On the contrary, it followed in the tradition begun by the illustrator-monks of the Middle Ages who had hidden images of their benefactors in their drawings of religious codices. Taurinus, too, whose historic poem about 1514 I've quoted earlier, opened his volume with a respectful acknowledgment to Margrave György Brandenburg, carefully inscribing many of his titles:

For His Royal Highness
Margrave György Brandenburg,
the worthy guardian of
His Majesty Lajos,
the mighty king of Hungary and Bohemia

It is unlikely that the British journal's copy editors would have insisted that Dirk write *Elizabeth A. M. Windsor* instead of *Queen Elizabeth II* or *Sayyid Abdullah II bin al-Hussein al Hashimi* instead of *king of Jordan*. The cabinetmaker, however, was insignificant enough to prove an exception.

Dirk wasn't ready to give up this fight, though, so he contacted the journal's editors, explaining Dennis's role in the discovery. Eventually, they sided with him. So, if you were to search the archives of *Nature* magazine, you will find that Dirk's paper on the laws of human motion ends with the following: "We thank cabinetmaker D. Derryberry for discussions and for drawing our attention to the wheresgeorge Web site."

In the pages prior to the acknowledgment, Dirk had documented at length the notable difference between the trajectory of the dollar bills and the random motion of Einstein's atoms. The distances atoms cover between two collisions are comparable, meaning that we never see a runaway molecule—one that goes a thousand times farther than the rest before it collides with another molecule. In contrast, while most banknotes resurfaced only a few miles from their previous locations, there were several truly runaway bills, traveling thousands of miles between two taggings.

When Dirk compiled the banknote trajectories, he realized that the distances covered by the bills did not follow the bell-shaped curve that Einstein had predicted, which would have meant that they all tended to travel roughly the same distance between sightings. Instead, Dirk found that they followed a power law, which, in this context, was uniquely significant: It indicated that the bills' trajectory is best described by what scientists have dubbed a *Lévy flight*. Similar to a person taking a wandering, aimless walk, a particle on a Lévy flight randomly changes its direction after each jump. That the jump sizes can be described by a power law means that most of the time a Lévy particle appears confined to a small area, tiptoeing around the same neighborhood. Occasionally, however, it will leap great distances before resuming its mincing dance in some faraway spot.*

Lévy flights were quite familiar to Dirk Brockmann and his mentor, Theo Geisel. Indeed, in the 1980s and '90s physicists had discovered that many seemingly random trajectories observed in nature, from tiny objects floating in turbulent liquids to the spread of matter in the universe, could not be described by Einstein's diffusion theory—meanderings of roughly equivalent distances, resulting in a fairly even dispersion of the particles. Instead, they followed a Lévy trajectory—huge leaps bookended by claustrophobic dithering. A few years earlier Dirk and Theo had found that our eyes also follow a Lévy flight when we explore a new image: First our attention wanders in the neighborhood of some detail, only to suddenly jump to a distant spot, surveying this new vicinity with many tiny eye movements.

But it wasn't the dollar bill–eye movement link that caught Dirk's attention. Rather, it was a decade-old study linking Lévy flights to birds and monkeys. Which is a rather remarkable connection, when you think

---

* What, you may ask, is the difference between a Lévy flight and a power law? Well, the power law relates to the Lévy flight just as the apple's taste relates to the apple itself. Indeed, random walks come in several flavors. If the randomly moving object covers the same distance during each jump (or if the jump sizes follow a Gaussian distribution), then we speak of a regular random walk. The atomic trajectories studied by Einstein belong to this class. There are objects, however, that move in a rather erratic fashion, such that their jumps follow a power law, like the dollar bills observed by Brockmann. As these trajectories have special properties, they have been given their own name—Lévy flights. Benoît Mandelbrot, the father of fractals, took the name from one of his mentors, Paul Lévy, a French mathematician. Thus, a Lévy flight is a special kind of random walk, and the power law is the property that distinguishes a Lévy flight from an ordinary random walk.

about it, as it hints that the burstiness we encountered before may not only have preceded the Internet, but its roots may underlie human will and consciousness.

∾

In 1995, between transatlantic flights, Sergey Buldyrev elected to spend his overnight layover in London visiting his second cousin, Vsevolod Afanasyev. Sergey thought of Afanasyev as the modern incarnation of the nineteenth-century Russian folk hero Lefty, a skilled blacksmith who had bolstered national pride. Legend has it that when the British offered Czar Alexander I a clockwork dancing flea—equal parts gift and implicit boast about the superiority of Western technology—Lefty ingeniously fitted the critters with minuscule horseshoes, each beautifully engraved with the artisan's signature. A triumph of Russian craftsmanship.

Like Lefty, Cousin Vselovod had an amazing gift for miniature devices. And also like Lefty, who had been sent by his czar to England, Cousin Vselovod had relocated to Cambridge to work for the British Antarctic Survey. He was not busy cobbling shoes for dancing fleas, however, but assembling tiny detectors that tracked the movement and flight of birds. One of his designs, for example, trailed the wandering albatross, a gracious animal whose ten-foot wingspan makes it the largest living bird on the planet. His detector was the first to offer evidence that albatrosses circle the globe without touching land, flying for months at a time above the roiling ocean.

On Sergey's visit to London "we drank a bottle of vodka and had a nice conversation," he recalls. As the vodka disappeared, the cousins' chat inevitably drifted toward their research. Sergey had spent a great deal of time in the past studying random walks, but his vantage point had always been Einstein's—namely, approaching the issue as an abstract mathematical problem. So his cousin's data on the seemingly random trajectory of an albatross piqued Sergey's interest, offering him a chance to compare the century-old random-walk theory to the flight patterns of real birds.

Armed with a Ph.D. in physics from Saint Petersburg State University, Sergey had immigrated to the United States six years earlier, in 1989, accompanied by his wife and their two toddlers. With scant understanding of the U.S. research establishment, he had set straight off for Boston University a mere two days after his arrival, in search of an author whose research publication he'd so admired back in the Soviet Union. To his shock, he found that the object of his professional admiration was neither a professor nor a doctor but a twenty-four-year-old graduate student lounging around his office in a T-shirt and shorts. As soon as Sergey recovered his equanimity and made his situation clear to Peter Pool, he found himself dragged down a neon-lit corridor to meet Pool's advisor, Gene Stanley. Sergey was already quite familiar with a string of influential discoveries Stanley had made concerning the structure of super-cooled water and phase transitions, a research field he had helped to create. As such, he never would have presumed to disturb the famous professor. But what he didn't know was that Stanley was also quite familiar with his personal situation.

Stanley had first visited the Soviet Union in 1973 as the organizer of a conference held in Moscow. Upon his arrival he was shocked to learn that three of the Russian physicists whose work he greatly respected had been banned from the meeting. Their crime? Applying for visas to emigrate to Israel. Ignoring his colleagues' warnings, Stanley breakfasted with the three refuseniks and later that day announced to the conference assembly that the three would indeed present their talks, but in a private apartment, and all were invited.

Before he could even finish his provocative communication, the microphone went dead, and he was left to shout out directions to the apartment as two plainclothes KGB agents dragged him from the room. Taken to the very top floor of the towering building and dragged before a large window, Stanley suddenly remembered the suspicious "suicides" of prominent dissidents in the Eastern Bloc. Yet he was to live another day. Eventually let go, he was even permitted to join the thirty-some scientists and handful of KGB agents in the small apartment for the underground session he had organized.

The experience left a deep impression on Stanley, and he soon became a champion of sorts for underrepresented groups and dissidents in science.* So when, fifteen years after Stanley's Moscow adventure, Sergey showed up in his office, the renowned professor-humanitarian realized in short order that the Russian before him was guilty of the same crime as his three colleagues back in 1973: He'd dared to leave the USSR to follow his Jewish wife. Ten minutes later, Sergey was hired. Three days in the United States, and he'd landed his first job.

Six years later, Sergey had come to see the albatross problem as an important challenge for Stanley's interdisciplinary research group. So, with Stanley's blessing, one of his grad students, Gandhi Viswanathan, began analyzing the time series of *wet* and *dry* signals sent from Cambridge by Sergey's cousin Vsevolod. The data had been collected from the small devices attached to the legs of albatrosses in the South Atlantic. A wet signal indicated that the bird's leg was underwater, while the dry signals corresponded to the flight times between their fishing breaks.

Gandhi soon noted that the signals were by no means regular: Instead, they telegraphed many brief intervals during which the albatrosses touched the water repeatedly. These wet bursts were followed by long dry stretches, suggesting that a bird flew a significant distance before finding another spot appropriate for fishing. Sound familiar? It should, because Gandhi's analysis indicated that the time between the consecutive wet signals followed a power law, implying that the albatrosses' fishing pattern was bursty, best described as a Lévy flight.

Biologists had long since noted that animals tend to forage for long periods in the same area before moving on to search for food in some far-off elsewhere. None of them yet suspected, however, that this haphazard pattern followed a precise law. When the Boston University group published its paper in *Nature* in 1996, showing just that, a flurry of activity among animal researchers followed. The novel mathematical formalism revived long-forgotten data sets, evidencing Lévy trajectories all over the animal kingdom, from the spider monkeys of the Yucatán Peninsula to reindeer, bumblebees, fruit flies, and gray seals. The paper showed that Lévy flights capture a universal mobility pattern, followed by most ani-

---

* In 2003 his efforts on the behalf of such politically and socially disenfranchised scholars earned him a Michelson Medal for Humanitarian Outreach from the American Physical Society.

mals during foraging. Which raised a simple but intriguing question: Why do animals behave that way?

~

By the time the albatross paper was published, Sergey had been at Boston University for seven years. Past forty, he was an oddball among the sea of twenty-something grad students. Profoundly knowledgeable about mathematical physics and willing to outwork even the most ambitious lab members, he was now an indispensable part of Stanley's excellent team. Yet as his peers settled into tenured professorships, instead of polishing his résumé Sergey chased an explanation for the mysterious Lévy patterns he saw in animal travel. Soon he identified the critical question he had to ask: How do you find a banana tree in the jungle?

If you have no idea where the tree is, you could search for it randomly: Start walking in some arbitrary direction, and if you find nothing, change direction after a while, hoping that you get luckier this next time around. The problem with such a random strategy is that you will likely cover the same region over and over, as a random walk is known to occasionally return to the same area by pure chance. You might avoid this redundancy by doing a methodical search instead, as the FBI does when it looks for evidence, cordoning the crime scene into strips and systematically sweeping them. The data from the animals indicated otherwise: The birds and the beasts follow neither Einstein's lead nor the FBI's protocol but the mysterious Lévy foraging pattern instead.

Then Sergey had a flashback to the times long past when he used to pick wild cranberries from the vast marshes of Russia.

"I realized that I was following Lévy flights myself," he recalls, "jumping from one tussock rich with the berries to the next, then after some time making long trips to distant areas of the marsh in search of better grounds."

Grad student Gandhi Viswanathan came through once again, and with help from Shlomo Havlin, one of the world's authorities on random walks, he translated Sergey's childhood experience into a theory. He showed that when foraging for a sparse supply of food, the best strategy, in fact, is a haphazard combination of many small steps and occasional long jumps.

The resulting paper introduced a new paradigm: Neither a regular nor a random strategy is optimal when hunting for scant patches of food scattered over a large area. The best methodology is to follow a bursty search pattern instead, as the long jumps will force you to explore different stretches, while the many short steps will help you uncover most of the food in your immediate vicinity.

The theory resonated well beyond animal research. In fact, scientists have since used it to address one of the prevailing mysteries of cell biology—how a transcription factor, a protein that governs the activity of our genes, figures out precisely where to attach to a DNA strand. The DNA's double helix is comprised of three billion nucleotide base pairs, so locating one particular base pair is akin to locating a single penny tossed somewhere along the 2,462-mile stretch of highway separating Los Angeles from New York. Miraculously, billions of cells in our body achieve this with ease. And the emerging consensus is that their success is aided by a bursty strategy—the protein attaches to the DNA at random and from that point searches via small steps in the neighborhood of the attachment point. If it fails to find the desired spot, it detaches and tries again in some other, distant location along the strand.

Today bursty search patterns explain an amazingly wide range of behavioral phenomena, from how people recall facts stored in their memory to how they locate information on the World Wide Web. In publication after publication, scientists have offered evidence that the most effective strategy for locating a given target is not the one that is the most obvious, systematic, and regular but a search strategy that is bursty, intermittent, and even haphazard.

~

In the light of these discoveries, Dirk Brockmann's paper on the dollar bills was far more than a piece detailing the Georgers' amusing hobby. It was immediately pertinent to an active research field, expanding the Lévy paradigm from albatrosses and monkeys to Homo sapiens. The dollar bills follow us, going where we carry and spend them, just as Cousin Vsevolod's detectors tracked the albatrosses. They revealed that humans, during their daily wanderings, follow the same bursty Lévy pattern that evolution implanted in our distant ancestors. So Dirk's finding had profound implications, demonstrating once again nature's frugality and its tendency to apply the same solutions in quite different settings.

Much of the media attention that followed Dirk's publication, however, failed to grasp just how unexpected his conclusions were. It wasn't puzzling that our human trajectory is similar to the monkeys' or albatrosses'. Rather, Dirk's finding challenged a theorem little known outside of specialized scientific circles, which is found in rarely read, dusty tomes expounding on the mathematics of random walks. The theorem predicts that if a particle follows a Lévy trajectory, the more time that passes, the farther the particle will drift from its release point. This means that the chance that a Lévy walker will return to a previously visited location diminishes with every passing second.

This is by no means a problem for an atom in a liquid or a gas, which, as far as anybody is concerned, can diffuse arbitrarily far. For humans, however, this theorem creates a paradox: If we *do* follow a Lévy flight, we rarely, if ever, will find our way back home.

As puzzling as it sounds, this may be the best explanation for György Székely's path. To the best of our knowledge, he started from his hometown of Dálnok, the Transylvanian village proud even today of its famous son. We also know that at some point he traveled more than three hundred miles to Belgrade, quite a distance in those days. But why would he have forfeited the comparative luxury and stability of his land and nobility for the perilous life of a mercenary? His two-hundred-mile trip from Belgrade to Buda poses yet another puzzle: With his purse now filled with gold, his bounty following his victory over Ali, why not return home?

So far, his trajectory is consistent with a Lévy path: many short journeys interrupted by a few long jumps that restart his life in some distant spot. Of course, in the light of Dirk Brockmann's findings, his voyage acquires a new meaning. But despite the statistical support for his permanently rootless state, there is still a human element that makes us wonder why György Székely couldn't find his way back home. The letter I found in the archives of Nagyszeben in 2007 was about to shed some light on this question.

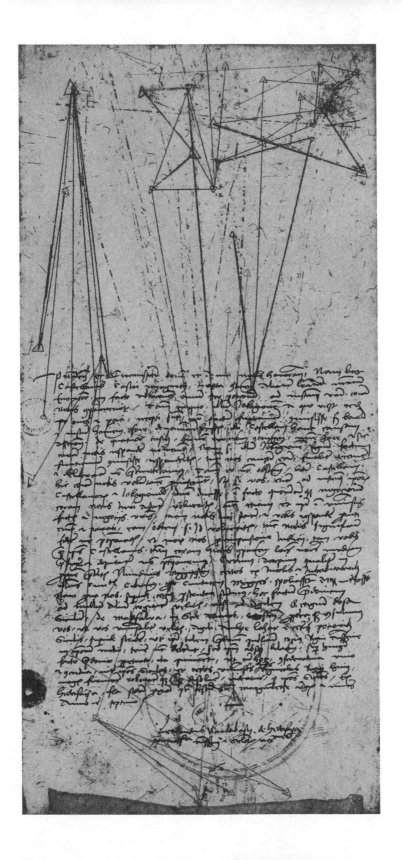

18

"Villain!"

July 19, 1507

t seems every family has at least one remarkable ancestor—a witty great-aunt whose personality shines through generations or a great-great-grandfather whose deeds survive the dustbins of history. My family went as far back as the sixteenth century to find our man. Lénárd Barlabási was, at the peak of his career, a Transylvanian vice voivode, next in line after Count János Szapolyai, who, as the voivode, governed the province.

It's not that Lénárd's descendants were undeserving. On the contrary, his cousin, János, was named bishop of Csanád, the episcopate vacated when György executed Miklós Csáky at Nagylak. Others are remembered for their tragic deaths, like Péter Barlabási, who was hanged when his patron, Prince György Rákóczi II, was deposed. Or there is Péter's son, István, who lost his life in battle after he joined Rákóczi on his frivolous campaign to claim the Polish throne. Many others on the leafy family tree may have been wonderful and accomplished men and women, but we know only of their quarrels. Much of what survived are court documents, detailing bitter fights among themselves or with others in the ongoing struggle to hold on to the wealth and influence Lénárd had established.

The life of Lénárd Barlabási is easier to trace because many of his letters have survived in various archives, turning him into an oft-quoted witness of the sixteenth-century social and political landscape of Transylvania. Unfortunately, his letters are not in Hungarian but the administrative Latin commonly used in those days. As a result they are indecipherable not only to me but puzzle even experts of medieval history. Indeed, Daniel Gregory Perett, a doctoral fellow in medieval studies at the University of Notre Dame, translated the 1507 letter I found in Nagyszeben into English for me but felt compelled to accompany the rendition with three pages of disclaimers and comments.

The first three paragraphs of Lénárd's 1507 letter discuss the case of a serf accused of theft. He was from Vingard, a small village near Nagyszeben. According to Daniel Perett's translation, the castellan of Vingard calls the serf a "good and just man" and categorically rejects the charges against him. Given all the he-said, she-said, Lénárd requested that all parties assemble before him at a future date.

The second half of the letter is especially interesting:

> Also, we have received news that on the most recent market day at Meggyes certain inhabitants and occupants of the township of Szeben near the township of Meggyes have been robbed and killed.
>
> We have discovered through careful inquiry that this act of banditry is to be attributed to none other than the knight György Dózsa, a Szekler of Makfalva located in Maros County.
>
> We therefore counsel you to write immediately to the knight András Lázár and the other Szekler lords, warning them that they should not harbor such a public robber of the whole realm of Transylvania in their midst but should rather punish him without delay for having committed such an act of banditry so that, as a result, it may be clearly seen among the Szeklers and all other good subjects of this kingdom that the

union and harmony of our confederation is being solidified and strengthened by the Szeklers, rather than weakened.

Given at Héderfája on the first Monday after the feast of the Blessed Virgin and martyr Margaret in the seventh year, etc.

Lénárd Barlabási of Héderfája,
Vice Voivode of Transylvania and
the Viscount of the Szeklers

Barlabási writes about a crime that took place in Meggyes, a Saxon town thirty miles from Nagyszeben. In writing "this act of banditry is to be attributed to none other than the knight György Dózsa, a Szekler of Makfalva located in Maros County," he introduces György Dózsa Székely for the first time in written history.

It remains something of a puzzle as to why the vice voivode did not offer to bring Székely to justice in the summer of 1507; in fact, he merely advised the lords of Nagyszeben to write András Lázár and the other Szekler lords to "punish him without delay." It may be the Szeklers' traditional independence that prompts his laissez-faire attitude. Whatever is behind his cautious approach, the letter played a key role in shaping the historical narrative of György's pre-Crusade career.

It is not that historians were particularly kind to György Dózsa Székely before 1876. The nobility and clergy—the chroniclers of those ages—not surprisingly treated with a combination of disgust and anger the man who dared defy them and the prevailing order. For example, József Eötvös's carefully documented 1847 historical novel depicts György Székely as an impulsive drunk, who, guided by the unfortunate combination of a fiery temper and uncouth advisors, sailed the winds of history in directions completely incomprehensible to him.

With the discovery of Barlabási's letter, György gains two additional epithets: robber and murderer. The document explains why he left Transylvania: He was chased by the vice voivode's order that the Szeklers "should not harbor such a public robber of the whole realm of Transylvania in their midst." So he joined the mercenaries of Belgrade, the dangerous no-man's-land, where any criminal could find shelter if he was willing to defend the border. According to this narrative, only seven years later, after he killed Ali in their duel, he felt secure enough to re-emerge at the Buda Castle. It is unlikely, however, that he went there anticipating any reward. He may have been every bit as astonished as the later historians when the court, short of heroes in the wake of an all-out war against the Ottomans, lavished him, an infamous outlaw, with honors and riches.

Not everybody was blinded by the praise, however. According to the popular narrative, György Székely was denied an audience by our prophet, István Telegdi, good counselor to the king of Hungary and opponent of the pope's Crusade. And, you will recall, Bishop Csáky, apparently aware of György's past, refused him the three hundred gold pieces the king had promised the knight as a reward for his brave act. The influential 1545 chronicle by Szerémi appears to know that "the bishop scolded him."

Károly Szabó, the Hungarian Historical Society member who first discovered the 1507 letter, is significantly responsible for the popular narrative. In the concluding paragraph of his 1876 report, he loses his normal professional detachment, expressing disgust over the turn of events: "Instead of punishing this ordinary robber and murderer, the court lavished him with rewards, honored him for his bravery, and Hungary's pontiff raised him to power, allowing him to unleash his and the mob's rage, leading the country to such a grave peril that its impact was felt for centuries to come."

Sándor Márki, György Dózsa Székely's biographer, echoed the same feeling in 1913: "It's unheard for the king and the cardinal to favor to such a great extent a man who yesterday was only a villain, a peasant, and a low-ranking officer and today is a famous nobleman and the commander-in-chief of a huge army."

Once Barlabási's letter surfaced, historians concluded that the dispute between Bishop Csáky and György Dózsa Székely must have been over the knight's past deeds. In this reading of the events, Csáky's execution at Nagylak was by no means a random act but revenge served cold.

Lénárd Barlabási and György Dózsa Székely were not the only names familiar to me in the 1507 letter. Lénárd's grandson, Bálint, acquired land in Ehed, a tiny village in the heart of the Szekler Land, where my grandfather, Albert, was born in 1909. By then the tongue-twister *l* was dropped from the family name, morphing it into *Barabási*. Orphaned early by a bullet that found my great-grandfather as he sauntered into the casino at Marosvásárhely on horseback, Albert was forced to leave Ehed. He eventually settled in Gyergyószentmiklós, one of the seven administrative centers of the Szekler Land.

A bike ride from my grandparents' house is Szárhegy, one of those villages whose name makes you careful with Hungarian accents. Write or pronounce it *Szarhegy* and the otherwise meaningless village name becomes *Shit Hill*. Quite the contrary, the village is the home of a beautiful Renaissance castle protected by tall stone walls and five towers decorated with exquisite hand-painted ornaments. Abandoned over the past century, its towers have sadly succumbed to the region's devastating winters, and the rocks of its once-strong walls have found their way into the foundations of nearby houses. Only the Franciscan monastery, built above the castle on the slopes of the Szármány Hill, remains largely undamaged. Even that august structure has suffered a slow deterioration since 1951, when the Communists outlawed the Franciscans and evicted the brothers. It was saved from complete abandonment, however, by local intellectuals who, in 1974, turned the monastery into an art center, hosting each summer about thirty artists from all over the country.

Given my intended career as a sculptor back then, at the end of my freshman year in high school I was sent by my parents to Szárhegy to attend the month-long art camp. Barely fifteen and without any specified duties, I did anything that needed doing: I carried the canvases of a famous elderly painter to the top of the mountain, posed for others, helped cut a window into a marble sculpture, and assisted in the preservation of the hand-painted Renaissance ornaments covering the castle's walls. By the end of the month I was intimately familiar with every rock of the magnificent ruin, including the infamous *kaszatömlöc*, a deep hole in which those condemned to death were dropped, to be torn to pieces by the sharp blades jutting into the falling convict's path.

While the castle's construction was finished only in 1632, its oldest tower dates back to 1490. That tower was built for the family of András Lázár of Szárhegy, the man into whose hands seventeen years later Lénárd placed the fate of György Dózsa Székely.

The joint appearance of György Dózsa Székely, András Lázár, and Lénárd Barlabási in the 1507 letter is perhaps a mere coincidence. Even so, it is a link that surfaces again when the fate of Transylvania is bestowed on these men.

19

The Patterns of Human Mobility

I am calling because I want to make an investment in you," said the man on the phone. He'd introduced himself as manager of a company I'd never heard of before. He spoke quickly—very quickly, in fact—leaving the impression that he had no time to waste. About a year after the publication of my first book on networks I had grown used to e-mails and calls from readers seeking advice on interconnected systems. This was one of the few times that someone had called not to ask but to give. He had my full attention.

The caller was a high-ranking executive at a mobile-phone consortium who'd recognized the value in having records of who is talking with whom. After reading *Linked* he had become convinced that social networking was essential to improving services for his consumers. So he offered access to their anonymized data in exchange for any insights our research group might provide.

His intuition proved correct: My group and I soon found the mobile users' behavioral patterns to be so deeply affected by the underlying social network that the executive ordered many of his company's business practices redesigned, from marketing to consumer retention. With that, he pioneered a trend that over the past few years has swept most mobile carriers, triggering an avalanche of research into mobile communications. Despite his crucial role in advancing network thinking in the mobile industry, his combination of modesty and caution prevented his ever wanting his name attached to any of it.

As my group and I immersed ourselves in the intricacies of mobile communications, we came to understand that mobile phones not only reveal who our friends are but also capture our whereabouts. Indeed, each time we make a call the carrier records the tower that communicates with our phone, effectively pinpointing our location. This information is not terribly accurate, as we could be anywhere within the tower's reception area, which can span tens of square miles. Furthermore, our location is usually recorded only when we use our phone, providing little information about our whereabouts between calls. Despite these constraints, the data offered an exceptional opportunity to explore the mobility of millions of individuals.

Marta González, a talented physicist from Venezuela, joined my group in 2006 fresh out of her Ph.D. program in Stuttgart, Germany, and took up human mobility as her main research topic. Given the huge amounts of data she had to sort through and the technical difficulties associated with legitimate privacy concerns, hers was a trying task. But she was determined to interpret the mountains of information, and her perseverance eventually paid off. She soon managed to extract a six-month record on the whereabouts of 100,000 anonymous individuals.

To our pleasant surprise, Marta's measurements were in excellent agreement with Dirk Brockmann's results: While most phone users traveled less than one to two miles between calls, occasionally some callers jumped hundreds of miles. Overall the distances followed approximately the same power law that Dirk had found when following the dollar bills. This was reassuring, confirming that humans, just like monkeys and albatrosses, follow a Lévy trajectory. But our celebration was short-lived. By the time Marta's results poured out of the computer, we were no longer quite so sure that it was a Lévy flight that we were looking at here.

Six years after the publication of their seminal paper, Sergey Buldyrev received fresh data on the flight patterns of the albatross from his cousin Vsevolod Afanagyev. Hoping that the more accurate data would bolster his 1996 findings, Sergey jumped to analyze it. To his dismay, he found that the long flights, the signatures of a Lévy flight, had vanished from the albatrosses' paths—the distances covered by the birds in the new data were mostly short and comparable to one another. It was as if the albatrosses had decided to intentionally follow a random walk, carefully avoiding a Lévy trajectory.

Puzzled, Sergey decided to reanalyze the data on which they had based their original *Nature* article. This time around, he noticed something odd: While the long flights were definitely there, they occurred mainly at the beginnings and ends of the birds' flight patterns. It looked as if an albatross started its day with a long flight, searching for a spot rich in squid, and then paused, repeatedly touching the water as it fed. Once satiated, it took off on yet another long flight back to its nest.

This, of course, made sense—what else would a tired albatross do who had a full belly and a cozy nest to return to? The problem was that, strictly speaking, if the albatrosses followed a Lévy trajectory, then the long stretches should have been dispersed more or less randomly throughout their flight histories, not limited to the beginnings and ends of their journeys.

Out of curiosity, Sergey eliminated the first and last flights from the original data, a move that should not have affected the statistics. But with that, the evidence for the Lévy flights vanished—the remaining path was simply a random walk.

While the origin of the long flights recorded at the end and the beginning of each bird's trip remained a mystery, they started to smell like an artifact of the data collection technique, not actual flights that the birds had taken. So, with great distress, Sergey started to suspect that the birds had not been following a Lévy path. Given the volume of research his 1996 *Nature* article had inspired, this new conclusion, if valid, would hit quite an array of scientific communities like a bombshell.

But the charge did not detonate. Not yet. It wasn't that Sergey had changed his mind; he remained convinced that something was not right. A critically busy spell in his life shifted his priorities, and the new findings sat in his office unpublished for years.

In the fourteen years he spent in Stanley's lab, Sergey had coauthored a staggering 190 papers, eight of which had been published in *Nature,* an output that many tenured professors would envy. While he loved his job at Boston University, his position was still only temporary, depending as much on Stanley's goodwill as on his ability to raise the necessary research funds to maintain Sergey's salary.

In 2004, about to turn fifty, Sergey concluded that time had arrived "to gain some respect," and he mailed out three applications for faculty positions at various universities. It was a modest beginning, and he knew that, given his late start, he would likely have to send out hundreds more before finding suitable employment. But just a few months later the improbable happened, and he was offered a professorship at Yeshiva University in New York City. He promptly packed and left Boston, just about the time he ought to have been finalizing the paper discussing his concerns on the Lévy flights. Burdened by the move and his heavy new teaching load, Sergey dropped the paper to the bottom of his to-do list, where it lingered for an astonishing four years, just as Theodor Kaluza's paper had languished on Einstein's desk. And it might have forever remained there if a new development hadn't once again reordered Sergey's priorities.

❧

At the end of October 2005 the British Atlantic Survey hired Andrew Edwards, a cheerful redhead with a degree in applied mathematics, to serve as their uniquely titled biosphere-complexity analyst. Edwards soon developed an interest in the albatross flight patterns and before long noticed that the long jumps were mainly at the beginning and the end of the birds' journeys, unaware that Sergey himself had made the same finding only a few years earlier.

Edwards later wrote me that Richard Phillips, BAS's albatross biologist, had told him that "this would be due to the first and last flights not being modified to account for the time that the birds spent sitting on the nest." This correction was routinely performed in the more recent measurements, like the new data set Vsevelod had sent his cousin in 2002.

As he dug deeper, Edwards learned that several birds in the original data set had carried a rudimentary satellite transmitter that had recorded their locations as well. Once he located the files, Phillips's suspicion was confirmed: During the puzzlingly long dry periods at the beginnings and ends of the recordings, the birds were not actually flying. Rather, they were sitting in their nests, snug and dry. Once these rest periods were removed from the analyses, Edwards realized, as Sergey had, the newly calculated flight pattern was indistinguishable from Einstein's atomic trajectories. A simple random walk, that was all he had evidence for in his computer.

In 2007, eleven years after its initial discovery, the Lévy character of animal foraging was no longer considered a hypothesis but a well-established scientific fact and had inspired hundreds of publications by ecologists, animal researchers, mathematicians, and physicists. Thus the entire scientific community was shocked when they read that year's October 25 issue of *Nature*: a paper coauthored by Edwards, Sergey, and many others concluded that any likeness the path of the wandering albatrosses had to a Lévy trajectory was an artifact of the measurements.

The news reached us just as Marta was finalizing her analysis of the cell-phone data, which had been suggesting that humans, like albatrosses and monkeys, follow a Lévy path. Suddenly, however, the Lévy paradigm was in shambles, prompting *Science* magazine to run a story asking, "Do Wandering Albatrosses Care about Math?" The controversy only sharpened our attention, forcing us to carefully inspect the validity of the Lévy paradigm for humans as well. Sure enough, we were in for some surprises.

❧

To appreciate the importance of the mobile-phone data, we must realize that the Georgers track our *dollar bills,* not us. That is, the reemergence in Florida of a bill stamped by Gary at the Ohio gun show by no means implies that the fellow who bought gunpowder from Gary is now off hunting for crocodiles in the Everglades. Chances are that he spent the bill at once, paying for a bumper sticker at the next booth. Then the bill may have been returned to an amateur hunter, who used it at a Pennsylvania gas station, where it soon saw itself in the pocket of a Florida-bound semi driver. By the time somebody in Miami Beach took the time to enter the bill's serial number at WheresGeorge.com, the banknote may have been owned by dozens of individuals who were by that point spread all over the world. Therefore, the jumps of the banknotes do not reflect the movement of any single person. Rather, the bills are like the baton in a relay—they go from runner to runner, capturing the overlapping journeys of several individuals.

Given these limitations, an important question eluded Dirk: how far people actually travel on a daily basis. Dirk could see the traces of the dollar bills, but not the individuals who carried them. It was as if you were watching a relay in the dark, where only the baton was emitting light, moving mysteriously along the track.

Marta, with access to our mobile phones, had a much more penetrating microscope, seeing in real time everyone's movements. Working together with Cesar Hidalgo, a graduate student in my group, she reconstructed a user's trajectory and then drew a circle around it, sizing up the neighborhoods the user frequented. She then proceeded to do the same for a hundred thousand users.

Marta's ability to draw these circles worried us, as in theory no individual or monkey following a true Lévy trajectory should be confined by any boundary. Indeed, as soon as you establish a monkey's past whereabouts, it is bound to jump away in search of a banana tree in some previously unexplored spot. In general, the longer the monkey follows a Lévy trajectory, the farther it will drift from its starting point, thus necessitating an ever-larger circle to map the territory it has roamed. Remember the theorem predicting that if humans should follow a Lévy trajectory they would rarely, if ever, return home? They would drift farther and farther away, until they'd eventually die.

As predictive as the theorem was, humans obviously were not aware of it. Indeed, Marta found that instead of drifting toward exotic distant places, most of an individual's life is confined within a stable circle. That is, each of us tends to greatly limit our mobility to a few well-delineated locations.

Once again, none of us found this particularly surprising—we have a home and a workplace, and we spend most of our lives shuttling between the two spots and scarcely beyond. The main reason Marta's findings interested us is because they were so clearly at odds with the Lévy predictions.

The real surprise came when Marta compared the radii assigned to each user. How different are the distances I cover in my daily roaming from those you and thousands of others travel? she asked, exploring the heterogeneity inherent in the population at large. Obviously, unless we live and work together, you and I move in different corners of the country. But given the drag of commuting and the thrill of discovery that affects all of us similarly, shouldn't the size of our circles be comparable?

Not really: Once again a power law emerged from Marta's analysis. It indicated that most of us confine our lives to a very small circle, a few miles at most, moving back and forth with high regularity among several nearby locations. This highly localized majority coexists with some people who move dozens of miles each day, and a few individuals who travel more than hundreds of miles. These are not people who occasionally take a trip, as we do when we travel for vacation or business. They are like Hasan Elahi, whose "itchy feet" drove him to routinely crisscross continents.*

Hasan does have a home and a workplace, but what is strange about him is that it is almost impossible to find him in their vicinity. So when it comes to the distances he regularly covers, compared to most individuals he does not appear normal. He is an outlier.

While his outlier status did make him special, he was by no means

* It's not that there are two groups of users, a group of numerous individuals who travel little and a smaller group of people who cover hundreds of miles on a daily basis. The power law describes a continuum between the two groups: There is a very large number of individuals who truly confine their mobility to a neighborhood of two miles or less. There are fewer, but still many, whose radius is about ten miles, and even fewer who move about fifty miles daily. These folks coexist with an even smaller group of people who are spread widely, covering hundreds of miles. In this respect, Richardson's formulation is appropriate: the larger, the fewer. The farther you travel regularly, the fewer individuals you will find like you.

unique. It turned out that there were many others just like him in our database. If all of us covered comparable distances on a daily basis, outliers would be extremely rare and thus surprising. In fact, they are effectively forbidden in a world where Poisson or Gaussian distributions describe our travel. In that universe my circle would be statistically indistinguishable from your circle. Yet, outliers are not only tolerated but also expected when power laws dominate our daily mobility. They are the equivalents of the world wars in Richardson's war database or the Rockefellers and the Gateses in Pareto's wealth distribution.

To be sure, Dirk was right: The bills *do* follow a Lévy path. What we could no longer conclude was that we do too. When pocketed by somebody who spends most of her time in a small neighborhood, the dollar bill won't get very far. But if the bill is passed to an outlier, like Hasan or a truck driver, it may travel thousands of miles before it is spent again. It is like a strange relay, in which many toddlers compete in the same team with a few Olympic athletes. When the toddlers have the baton, it moves randomly back and forth. Once an athlete takes over, you will have difficulty following its path.

At the end Marta discovered that the bills follow a Lévy path not because each of us follows a Lévy trajectory but because there are outliers among us. Remember, WheresGeorge tracked the currency, not the consumers. Society's heterogeneity—the huge differences between the many homebodies and the few globetrotters—made the bills stay put for a long time in the same neighborhood and occasionally jump across the continent. A strange but natural pattern—one that lay at the heart of Dirk's original discovery.

~

Four months after Edwards, Sergey, and their teams rebutted Lévy patterns in albatross foraging, *Nature* published yet another take on the subject. At the Marine Biological Association Laboratory in Plymouth, England, animal researcher David Sims had assembled a large group of researchers interested in marine animals and their mobility. Together his team collected millions of data points pertaining to the motion of several aquatic species. They concluded that sharks, bony fish, sea turtles, and penguins all exhibit Lévy flight–like travel patterns. With that, the study of Lévy flight has become akin to taking a nauseating roller-coaster ride.

But let's retrace its tortuous path just one more time. In 1996 Sergey and his collaborators reported that albatrosses follow a Lévy trajectory. The finding inspired an avalanche of research concluding that a wide range of animal species, from monkeys to bumblebees, all move following Lévy patterns. Then in 2007 Sergey and his collaborators took a step back, revealing that the corrected trajectory of a wandering albatross is consistent with a random walk. Yet if a Lévy flight offers the best search strategy, why didn't natural selection force animals to exploit it? In February 2008 David Sims showed that it did, in fact. So despite the problems with the data recording, which was beyond their control, the paradigm established originally by Sergey is probably the correct one and we are back to square one: While we are not sure what albatrosses do, most animals do actually follow a bursty Lévy path.

Science itself often follows a Lévy pattern—a huge jump ahead is trailed by many small, localized steps that appear to take us nowhere, or perhaps even backward in some instances. These are not wasted moves, however, but necessary to testing the boundaries of the new paradigm.

When it comes to humans, the twists in the story are equally fascinating. In 2006 Brockmann found that dollar bills follow a Lévy trajectory, suggesting that when it comes to our daily wanderings we are not so different from albatrosses or monkeys. The finding made lots of sense, illustrating how hard it is to escape the foraging habits that evolution has hardwired into our brains despite the fact that locating scattered sources of food is no longer crucial to survival. Yet when mobile-phone records allowed us to track individuals anonymously, we learned that the dollar bills' jumps reflect not any given person's travel patterns but the differences between our travel habits instead. The widely different jumps of the dollar bills capture an extreme population heterogeneity that affects everything, from the spread of the viruses to resource management in cities.

The truth is that you and I do not drift over large distances, as a Lévy particle does or a monkey does when foraging for food. Instead, anywhere we go, we soon return to home. Boring, you might say, and it might be so—but I hope you stay tuned to witness the conclusion. Because while you and I may appear to be completely normal, we have Hasan and a few other outliers, with outsized appetites for travel, who make the world quite fascinating.

In the end, the mobile-phone data helped us resolve this paradox: If individuals follow a Lévy trajectory, they should never find their way home, and yet they do. The prediction is certainly valid but applies only if we follow a Lévy trajectory. In the meantime we established that we do not wander eternally homeless, which releases us from the theorem's vice grip, allowing us to go home again. Which takes us back two chapters: Does this mean that György Székely—disgraced, honored, and imperiled in succession—was finally allowed to return home, as we all do, at the end of the day?

Given his decision to turn the Crusades into an uprising against the aristocracy, the cardinal and the pope no longer called the shots. So, as his host's supreme captain he was now free to do anything he wanted. It was only a matter of priorities. If he so chose, he could even head back to the Szekler land. There was one caveat that shadowed any future ambition, however: Wherever he went, he was followed by an army of men who now pinned their hopes on him. This, naturally, made him think twice before hitting the road. Really, where would *you* go if thirty thousand fans insisted on tagging along?

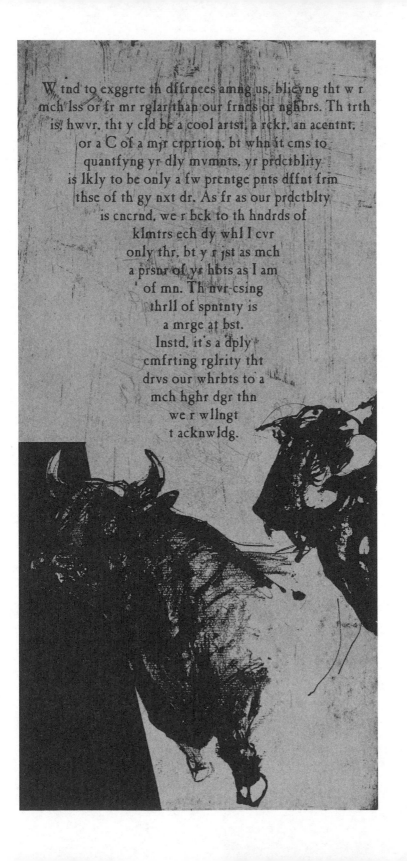

W tnd to exggrte th dffrnces amng us, blieyng tht w r
mch lss or fr mr rglar than our frnds or nghbrs. Th trth
is, hwvr, tht y cld be a cool artst, a rckr, an acentnt,
or a C of a mjr crprtion, bt whn it cms to
quantfyng yr dly mvmnts, yr prdctblity
is lkly to be only a fw prcentge pnts dffnt frm
thse of th gy nxt dr. As fr as our prdctblty
is cncrnd, we r bck to th hndrds of
klmtrs ech dy whl I cvr
only thr, bt y r jst as mch
a prsnr of yr hbts as I am
of mn. Th nvr-csing
thrll of spntnty is
a mrge at bst.
Instd, it's a dply
cmfrting rglrity tht
drvs our whrbts to a
mch hghr dgr thn
we r wllngt
t acknwldg.

20

Revolution Now

Temesvár, summer 1514, little more than a month after the May massacre

ifficult to storm and easy to defend, the fortress of Temesvár was second only to Belgrade in importance where Hungary's defenses were concerned. Its strength lay as much in its inaccessibility as in its fortifications. From the south and east it was protected by reedy swampland that proved equally impenetrable on foot and by horse. And from the west the River Bega acted as a natural moat, which left only a narrow strip of land connecting the fortress to the outside world. A system of planks, thick walls, and square stone towers built by János Hunyadi more than half a century earlier allowed the defending archers and musketeers to secure each rock on the town's perimeter. When Hunyadi departed in 1456 to liberate Belgrade from the Ottoman siege, he left behind a fortress that could easily withstand the cannonballs that György Dózsa Székely's crusaders were now shooting at it.

Despite its strength, or because of it, Temesvár had to be taken. It harbored István Báthory, whose assault on the outposters many blamed for turning the holy Crusade into a bloody civil war. For György Székely, however, the attack wasn't about revenge. Any army requires a fortress where it can retreat to safety if luck on the battlefield turns. He saw Temesvár, with its rich lands, strong walls, and prosperous merchants, as a satisfactory base for even an indefinite campaign.

Having barely escaped the sword after his defeat at Nagylak, Báthory never again underestimated György Székely's peasants. And so upon falling back to Temesvár, he wasted no time buttressing her defenses—patching walls, hauling food and armaments inside the walls, and hiring mercenary soldiers, going into debt to do so. The frenetic preparation paid off when on June 13 the defenders easily rebuffed the crusaders' hurried surge against the main gate.

The lives lost during György Dózsa Székely's unsuccessful attack were not entirely wasted, however, as the surge allowed him to gauge the strength of the defense. Taking the fortress by a direct attack would be a bloody business of uncertain outcome, he concluded, so instead he settled in for a long siege, chipping away the city's defenses, one cannonball at a time.

If the midsummer heat were to persist, he calculated, the swamps would dry up in a few weeks, allowing the crusaders to approach the fortress from the more vulnerable south and east sides. The Székely could not afford to gamble on the weather, however, so he gave thousands of his peasants a backbreaking task, to reshape nature's landscape. They began digging a five-mile canal that would divert the river from the moat and drain the surrounding wetland. Once the land was dry, the fortress would fall within days—a plan neither he nor Báthory doubted would succeed.

⁊

Before the Battle of Nagylak the crusaders' attacks on the holdings of the aristocracy had been isolated, and the few that had taken place were against György Székely's explicit orders. After the massacre, however, it was with their captain's blessing and leadership that the crusaders systematically occupied all fortresses along their path. By the time György's columns arrived before the walls of Temesvár, half of Hungary was under his command. Most cities and fortresses had opened their gates voluntarily, and the few that had refused were taken by force. Only the royal fortresses at Buda, Temesvár, and Belgrade successfully repelled his advance, and with the siege of Temesvár they hoped to change that.

During the lengthy siege György not only had to supervise the digging of the canal but also make certain that the king would not come to Báthory's aid. His strategy was simple: Occupy the nobility throughout the land with their own troubles, preventing them from forming a potent liberating army. To that end he sent a unit led by his younger brother, Gergely, westward toward Buda. En route Gergely conquered Csanád, the center of the late Bishop Csáky's episcopate, and several other cities. Toward the northeast, one of the company's most celebrated captains, Brother Lőrincz, marched his men toward Nagyvárad, an important fortress at the threshold to Transylvania. Finally, the Serbian Radoszlav and his forces turned their attention south, subduing much of the rest of Hungary, all the way down to Belgrade.

The crusaders received unintended help from the king himself, who asked János Bebek, a prominent knight and aristocrat, to lead the nobility's campaign against the peasants. The impatient Bebek failed to wait for a full army to assemble, an especially protracted process now that the nobility was reluctant to leave their families to the mercy of the marauding peasants. He foolishly attacked a Crusade unit with insufficient forces, only to be promptly and thoroughly defeated.

As word of Bebek's failure reached the court at Buda, the panicked king named another aristocrat, István Perényi, to helm the nobility's forces but failed to first rescind leadership from the disgraced Bebek. Soon Hungary had two rival noble armies that spent weeks squabbling over each other's legitimacy, leaving Temesvár to fend for itself.

Not even three months had passed since the cardinal had named György Dózsa Székely commander of the Crusades on April 24. In this short period of time the exiled mercenary had transformed into an inspiring leader, combining brilliant strategy and disciplined execution to conquer much of Hungary. Never mind that generations of historians would eventually label him a robber and murderer; military strategists today continue to admire the way György Székely planned and conducted Temesvár's siege. Even those disapproving historians who condemn György Székely agree that what was often depicted as an incoherent peasant uprising was in reality a systematic all-out war, expertly executed and strategized by an astute commander-in-chief.

In addition to deftly handling a war on many fronts, György was re-engineering the social order he had inherited, infusing the culture of war with Szekler tradition. It is customary to credit the Franciscan monks and friars present in the camps for much of the ideological shift that occurred in 1514. The truth is, however, that the methods adopted by the crusaders can only be understood from the unique cultural perspective György and his brother, Gergely, brought from the Szekler Land. As a contemporary Venetian observer noted, the crusaders were aiming for no less than *renovar el Regno de Hungaria,* leaving only the king, a bishop, and two lords at the country's helm.* This was in keeping with Szekler law, which acknowledged no other authority than that of the king, the *ispán* (bailiff), and the *királybíró* (king's local representative).

Additionally, György Székely had promised to redistribute the aristocracy's land among his serfs and peasants in proportion to their roles in the Crusades, once again in deference to a Szekler custom that assigned communally owned land to individuals according to military rank. When, in the Proclamation of Cegléd, Székely called the citizens of Pest and Szolnok to arms, he was not asking for volunteers but commanding every man to join his forces. Those who refused were assured that György Székely "[would] have their houses ruined, and we [would] not pardon their families either." This was an unusually severe punishment without Hungarian precedent. Yet the decree would have been familiar to the Szeklers, who ritually demolished the houses of those among them who neglected their military duty.

How did an ordinary mercenary, one whose brightest pre-Crusade moment had been the robbing of a merchant caravan seven years earlier, turn into such an inspired leader? If in 1507 he was truly the petty criminal he is so often depicted to have been, why wasn't he arrested by the vice voivode, Lénárd Barlabási? Politics.

---

* *Reform or renew the Hungarian kingdom.*

Many history books fail to note that the year preceeding Lénárd's 1507 letter was a particularly turbulent time in Szekler Land. Noble by birth, the Szeklers were exempt from taxes in exchange for military service. Only the ox tax was expected of them, the nominally voluntary donation of one ox per household on the occasion of the king's coronation, his wedding, or the birth of his firstborn son.

In keeping with the tradition, the Szeklers had delivered more than ten thousand oxen when King Ulászló was crowned in 1490 and yet again in 1502 on the occasion of his wedding. When the heir to the throne, the future Luis II, was born on July 1, 1506, the king once again expected the gifts. This time, however, the Szeklers offered nothing, reasoning that the donation was due only if the king's *firstborn* child had been male, which was not the case.

Nevertheless, the cash-strapped king sent tax collectors to seize his tribute, and the Szeklers, hardly known for their docile nature, promptly killed them. The embarrassed king now dispatched Pál Tomori, captain of the fortress at Fogaras, to restore order. Despite writing that the Szeklers "are frightened by neither God nor man," Tomori and his cavalry of five hundred approached the mission with confidence, knowing that though the Szeklers were quick to anger, they also held the king in deep respect. Indeed, never before had this people turned their swords against the crown. And so it was to Tomori's great surprise that his cavalry was actually met near Marosvásárhely by the assembled Szekler forces and thoroughly beaten. Tomori himself, bleeding from twenty wounds, barely escaped with his life.

Once news of the fresh defeat reached the court, the king ordered an even larger force assembled. He now called on the Saxons of Nagyszeben to subdue the disobedient Szeklers. By the time the Saxons arrived, however, the Szeklers had returned to their homes and decided that if the king was in such dire need of cattle, they might as well give them to him.

Only a year had passed since this blood had been spilled, and the Szeklers had not forgotten how the Saxons had joined the king's campaign against them, raiding their lands and killing several Szekler leaders in the process. The resulting bad blood undermined the union of the three

Transylvanian nations—the Hungarians, the Szeklers, and the Saxons. To restore the balance of power, the Szeklers felt that they needed to teach their neighbors a lesson. The three-volume *History of Transylvania,* written in the 1980s, which offers by far the most detailed historical account of the region, records that "the next year's attack on the Saxons of Nagyszeben, who had participated in suppressing the Szekler's uprising, was led by György Dózsa from Makfalva, who is reputed to have become a mercenary at the Ends under the name György Székely."

So Barlabási's 1507 letter that I read in the archives was not really about a robbery, as its literal reading suggests. As viscount of the Szeklers, Barlabási must have been aware of György Dózsa's role in settling his nation's score with the neighboring Saxons. But as vice voivode of Transylvania, he had to rise above the infighting of the three nations, making sure that "the union and harmony of our confederation is . . . solidified and strengthened." So he intentionally labeled György Dózsa Székely a criminal, termed the revenge a robbery, and, in so doing, astutely downplayed the gravity of the rebellion.

And so, in truth, György Dózsa Székely was not the petty robber and murderer he is made out to be by the 1507 letter. He came into maturity during a particularly politically charged time, when the jealously guarded freedoms of the Szekler nation were under threat. He must have participated in the 1506 battle that shattered the king's army near Marosvásárhely, as a year later he was entrusted by his peers with restoring the honor of his nation by avenging those the Saxons had slain during their unopposed raid on Szekler Land.

Indeed, who would have even remembered the death of a few merchants at the hand of a thief far away from the Buda Castle seven years after the fact? It was an event all too common in those days. Only György Székely's role in the political struggle against the king and his allies can explain why Bishop Csáky and János Telegdi were still aware in 1514 of Székely's not-so-recent past in Transylvania.

~

But back at Temesvár, under siege, Báthory was quick to realize the danger posed by the crusaders' ever-expanding canal. If his supplies held, he could defend the main gate indefinitely. It was much harder, however, to repulse an all-out attack directed at the weaker walls that bordered the swamps. Desperate to thwart György Dózsa Székely's plan, Báthory led a small cavalry unit in a surprise nighttime raid and massacred the peasants guarding the canal. Before the crusading forces could react, the newly built dam had been destroyed, pouring water back into the swamps.

This was a major setback for György, but he refused to be discouraged. He strengthened the canal's defenses, making sure that another breakout would be met with proper resistance, and carried on as before, believing that he had plenty of time to complete the siege. Redraining the swamps would now delay the final assault until August at least. By then, however, the month-and-a-half-long blockade would have exhausted the food supplies inside the fortress, meaning that his well-fed peasant forces would have even less trouble overpowering defenders weak with hunger.

Meanwhile, György's splinter armies having caused far-reaching disarray among the nobles, he knew he had little to fear from them. Only one person still had the power to confront him: Count János Szapolyai, Transylvanian voivode. It was an open secret, however, that the voivode hated Báthory just as much as the crusaders did, so he was possibly the last person expected to aid the fortress. Besides, the last time the crusaders had heard word of Szapolyai, he had been in the south, still waiting for the peasant army to join his march to Constantinople.

György Székely had no illusions of safety, however; by now Szapolyai must have learned about the revolution that had engulfed Hungary and was bound to return. So the crusaders readied a proper welcoming party: Brother Lőrincz left Nagyvárad, taking his army straight to the heart of Transylvania. Swollen with fresh recruits, his forces aimed for Kolozsvár, one of Transylvania's most prosperous cities.

If the voivode wanted to keep a grip on his country, he had no choice but to confront Brother Lőrincz's growing army. And that would offer precisely the delay György Dózsa Székely needed to finish his siege. And once György had taken the Temesvár fortress, a new balance of power would reign. At that point the king and voivode would have had no choice but to treat György as their peer.

21
Predictably Unpredictable

Since the publication in 2005 of *The Traveler*, a New Age "high-tech paranoid-schizophrenic thriller" with an Orwellian twist, a peculiar debate has absorbed cyberspace. The book takes us into a world where life is free of crises and surprises, a world of ennui-inducing normality. This peace and apparent security is maintained by a worldwide system of computers called the Vast Machine, fed by millions of surveillance cameras, sensors, and detectors. Only the members of a once-powerful ancient society and their sword-carrying protectors, the Harlequins, are aware of the Vast Machine's reach and are willing to stand up to it.

The ongoing debate this book continues to inspire on blogs and bulletin boards alike might easily focus on the eerie parallels between our own post-9/11 society and the tightly monitored world described in it. But it does not. It might also center on the book's literary merits, except that, as one critic put it, the writing "is pitched to perhaps a seventh-grade reading level," an assessment few would challenge. The debate is instead about John Twelve Hawks, its author.

The blockbuster sales and movie rights ought to have elevated Hawks to national celebrity, putting him among the likes of Stephen King and Dan Brown. Yet they did not. And it isn't because the media shuns him either. The real reason that you never hear about Hawks is that nobody seems to know him. He does not sign books and does not participate in promotional tours. In fact, he has never been seen in public and supposedly communicates even with his editor only through an untraceable satellite phone. Just like the Harlequins on perpetual run from the Vast Machine, John Twelve Hawks *lives off the grid,* a paranoid seclusion that fuels ongoing speculation regarding his true identity.

The book's central character is a Harlequin who preserves her off-the-grid anonymity by never using credit cards, opening bank accounts, or staying at permanent addresses. Aware that "any habitual action that showed a Harlequin taking a daily, predictable route to some location" will allow the Vast Machine to predict her whereabouts, she "cultivates randomness." That is, she relies on a random-number generator, or RNG, to guide her decisions. "An odd number might mean Yes, an even number No. Push a button, and the RNG will tell you which door to enter," freeing her actions from predictable patterns.

The book is a tale of a battle between good and evil that takes us briefly into something like that fifth dimension Theodor Kaluza proposed to Einstein, throwing into the mix Japanese sword fights and quantum computing. It also again begs the question, could one build a Vast Machine that foresees our actions?

We find it perfectly acceptable that particle physicists can predict within a picometer of accuracy the trajectory of a proton or that rocket scientists can launch a satellite that nine months later drops a robot on Mars. Unlike protons or satellites, however, humans tend to seek new experiences in a continually changing world, making it impossible to foresee their long-term actions. Indeed, given my hectic travel schedule, until recently I found any attempt to predict my whereabouts a few weeks in advance to be a hopeless exercise, fueling my hope that the Vast Machine will always stay where it belongs—in the realm of science fiction. Lately, however, I have begun to have my doubts.

~

There are few stereotypes so deeply ingrained in our collective psyche as the belief that young equals wild and unpredictable. Fueled by memories of the disestablishmentarianism of the Western 1960s and '70s counterculture, along with the speeding trendsetter economy of today's cyber youth, "Young and wild" is the implicit slogan of countless advertisement campaigns, movie scripts, and top-forty hits. As a result we tend to romanticize college life, the cradle of youth culture, seeing students as perhaps the most spontaneous and thus least predictable segment of the population. Yet Sandy Pentland, an MIT professor who follows the chatter of hundreds of students every day, finds that concept preposterous.

In the early 1990s Pentland started a research program in wearable computing at the Media Lab at MIT, prompted by the realization that, given the rate at which computers were shrinking, we soon would want to have them with us all the time. Sandy's vision of the future proved remarkably accurate, as today computers have become a part of our wardrobe, fashion accessories of a kind. In fact, for the most part we have stopped even calling them *computers*. We refer to them simply as *smart phones*.

In the fall of 2004 Nathan Eagle, a doctoral student in Sandy's lab, offered one hundred MIT students free Nokia smart phones, a desirable top-of-the-line gadget at the time. This was no handout, however; the catch was that the phones collected everything they could about their owners: whom they called and when, how long they chatted, where they were, and who was nearby. By the end of the year-long experiment, Nathan Eagle and Sandy Pentland had collected about 450,000 hours of data on the communication, whereabouts, and behavior of seventy-five Media Lab faculty and students and twenty-five freshmen from MIT's Sloan School of Management.

Trying to make sense of his data, Nathan arranged each student's whereabouts into three groups: home, work, and "elsewhere," the latter category assigned when they were neither at home nor at work but jogging along the Charles River or partying at a friend's house. Then he developed an algorithm to detect repetitive patterns, quickly discovering that on weekdays the students were mainly at home between the hours of ten P.M. and seven A.M. and at the university between ten A.M. and eight P.M. Their behavior changed slightly only during the weekends, when they showed an inclination to stay home as late as ten A.M.

None of these patterns would shock anybody familiar with graduate-student life. But the level of predictability of their routines was still remarkable. Nathan found that if he knew a business-school student's morning location he could predict with 90 percent accuracy the student's afternoon whereabouts. And for Media Lab students, the algorithm did even better, predicting their whereabouts 96 percent of the time.

It is tempting to see life as a crusade against randomness, a yearning for a safe, ordered existence. If so, the students excelled at it, ignoring the roll of the dice day after day. Indeed, Nathan's algorithm failed to predict their whereabouts only twice a week, during rare hours of rebellion when they finally lived up to our expectation that they be wild and spontaneous. Yet the timing of these unpredictable moments was by no means random—they were the typical party times, the Friday and Saturday nights. The rest of the week, twenty-two out of twenty-four hours a day, the students were neither the elusive Osama bin Laden nor the ubiquitously erratic Britney Spears but instead dutifully trod the deeply worn grooves of their lives. So maybe the Harlequins were onto something when they insisted on using an RNG. Had they studied at MIT, their whereabouts would have been no mystery—not to Nathan, nor to the Vast Machine.

But we may yet avert the dawn of an Orwellian world as described in *The Traveler*. For me, this sense of hopefulness emerged in the summer of 2007 when I purchased a brick-sized wristwatch. It was a loud antifashion statement and doubled as a GPS device, which recorded my precise location every few seconds. After I had worn it for several months, Zehui Qu, a visiting computer-science student, applied Nathan Eagle and Sandy Pentland's predictive algorithm to the data collected by my GPS. Sure enough, after a few days of training, Qu was able to predict my whereabouts with 80 percent accuracy.

While the algorithm's performance was impressive, the persistent gap between the 96 percent predictability Nathan found among the MIT students and my 80 percent raised a red flag. Neither I nor the MIT students were a fair representation of the population at large. Marta's study of the mobile-phone records had already explained why: When it comes to our travel patterns, we are hugely different. Some, like the MIT students and myself, are relatively home- and office-bound. Others are outliers, however, and travel a lot, tending to be less localized.

So does that mean there are people out there who are far less predictable than the MIT students and I? Truck drivers, perhaps, who travel the country for weeks at a time? Soccer moms, whose minivans shuttle between piano and fencing lessons? What about supertraveler Hasan Elahi, whose "suspicious movements" will undoubtedly land him in hot water again? How different are they from you and me? Are there Harlequins among us, individuals whose lives are driven by the roll of the dice to such a degree that their movements are impossible to foresee?

∝

If our make-believe friend Daniel goes to work at eight A.M. each weekday, has lunch at the same cafeteria, always at noon, and leaves the office for home promptly at six P.M., where he stays until the following morning, there is little mystery as to his future whereabouts. In the language of physics or information science, Daniel's entropy is zero. In other words, he is fully predictable.* In contrast, the Harlequin using an RNG will have a maximum entropy, making her whereabouts impossible to foresee.

If I want to know how predictable you are, I need to first determine your entropy, which is exactly what Chaoming Song attempted to do, not for one, but for countless individuals. Chaoming, a bright postdoctoral research associate who joined my lab in the spring of 2008, examined the data from the millions of users in our mobile-phone database but quickly learned that determining each caller's entropy was much easier said than done.

---

* *Entropy* measures the degree of disorder (or the lack thereof, which would be the degree of *order*) characterizing a system. It was Ludwig Boltzmann, an Austrian physicist, who connected the entropy, S, and the number of states available to the system, Ω, via the formula $S = \log \Omega$. In other words, if there is no ambiguity in a system's current status, then only one state is available to it, that is, $\Omega = 1$, and thus, the entropy is zero. A system, however, that is equally likely to be found in N different states will have a maximum entropy $S_{max} = \log N$. If Daniel has a truly regular daily pattern, there is no ambiguity as to his whereabouts at any given moment. Thus, for him, $S = 0$. For a Harlequin, however, equally likely to be at any of the N different locations, the entropy is $\log N$. Boltzmann's contemporaries considered the entropy so fundamental that they carved it on his tombstone. They were right.

Song's biggest difficulty was that most of the time he had no clue where the users were. Indeed, mobile-phone carriers record our location only when we actually are on a call. Our call pattern is bursty, however, which means that there are short periods with multiple location readings, when we make many calls in succession, and long intervals when no location is registered, because we are not using our phone. This haphazard record makes each user appear less predictable than he or she really is. Indeed, Daniel might move with clockwork precision among his favorite spots—home, office, and cafeteria—but if we knew his whereabouts only when he made phone calls, it would take quite a while to discover how regular he really was. If he occasionally were to deviate from his routine, skipping lunch for a stroll in the nearby park or taking off early to join his friends for a nacho grande at the nearby pub, with the scattered data points we have on his location, his whereabouts would surely appear to us to be quite random. To a certain extent, I found this comforting, since it suggests that bursts shroud us in such a way that we are hard to track and even harder to predict. I soon learned, however, that our burstiness does not keep us off the radar screen altogether.

Chaoming Song got an unexpected hand from an important property of our daily activity: redundancy.* If we were to go on an exotic vacation, our friends might enjoy receiving regular updates about our doings and whereabouts. On a regular workday, however, calling them on an hourly basis to update them on our goings-ons would soon grow terrifically annoying. Indeed, after a while our reports would cease to provide any new information at all, becoming a droning repitition: I am at work. Working. Still at work. Again with the work. *I know, honey—you are always at work.*

Chaoming Song was able to reap the benefits of this redundancy,

* The best way to illustrate redundancy is by means of an example proposed by Claude Shannon in his landmark 1948 article, which has become the founding document in information theory. Using the concept of entropy he showed that English has a 50 percent redundancy, which means that about hlf of th ltrs o ths txt ca be dletd and w cold stll dcipr is mnig. A redundancy-free language means that all letter combinations are meaningful, so a missing or mistyped letter would give a completely different meaning to any word, like *walking* or *waking,* or *goddess* or *godless.* In reality, while THE has a precise meaning in English, HTE, EHT, TEH, or ETH do not, allowing us to misspell THE with abandon while retaining its legibility. This is why making a good puzzle is so challenging—if we place words along the horizontal axis on a grid, the combinations they spell out on the vertical axis are often meaningless.

which allowed him to do what had initially seemed impossible: removing the veil of uncertainty that the burstiness of the call pattern had placed over us. That is, relying on the deeply repetitive nature of our habits, Chaoming cleverly developed a procedure that accurately estimated each user's real entropy. With that we could finally deliver a quantitative answer to the question that's been pushing me all along: Just how predictable are we?

❦

In 1927 the young German physicist Werner Heisenberg discovered an inequality that today is widely known as the *uncertainty principle*. It told us in no uncertain terms that the more you know where an object is, the less you know where it goes. That is, if we strive to know the precise location of a particle, we would be inherently uncertain about its velocity. If, however, we somehow measure its speed, we will not know for sure where it is.

What made Heisenberg's prediction counterintuitive is that it had nothing to do with the quality of our measurement—it told us that even the best experimenters would be unable to simultaneously determine the precise location and speed of a particle. The result was thus fundamental, valid to all objects, from electrons to humans. To be sure, for a moving bike or a rapidly approaching car the predicted uncertainty is so minimal that none of us would ever notice it. Yet it is there. (An overly close brush between a certain bike and an approaching car did once result in the breaking of my own wrists, but Heisenberg can hardly be faulted for that.)

In the spirit of Heisenberg's uncertainty principle, I was wondering if there are fundamental limits to human predictability. Why can't I predict your future actions? Is it the inadequacy of my tools or the quality of the data that I have collected about you? Or is it possible that I am hitting some fundamental limit that I am not aware of? If such a limit does *not* exist, once we sharpen our tools and refine our data gathering, nothing that you do in the future will remain a mystery. On the other hand, if there *is* such a limit, discovering its precise nature is of profound importance, possibly revealing our absolute predictability, the degree to which our future can be foreseen.

Nick Blumm, a graduate student working in my lab, proved that such a limit *does* exist and that we are all subject to it. From my perspective, this was something of an ironic turn of events, considering that the finding came from someone with a résumé full of unexpected turns. After receiving a bachelor's in physics with high marks, Nick was poised for academia. But rather than following the standard track, he decided to change course. He taught English in Tokyo, earned a degree in mime from Marcel Marceau, tutored rich kids in Manhattan, and worked as the natural sciences curator for the Children's Museum in Brooklyn. After a decade of seemingly random wanderings though life, he read my book *Linked* and decided that he knew what he wanted. So, with that, he returned to school to get his Ph.D. in networks. As such, Nick was an ideal person to ask: Could anyone have predicted *his* unorthodox path? His answer? Maybe, but by no means guaranteed. To be precise, he proved that no matter how good our predictive algorithm is, for any user with entropy S, our tools are bound to make mistakes occasionally.

If Daniel's entropy is zero, in principle we could foresee his whereabouts with 100 percent accuracy. Most individuals, however, have nonzero entropy, which means that there is some degree of randomness in their mobility—they occasionally make turns that are impossible to foresee. Thus, each person has a maximum predictability such that no matter our efforts, we cannot be absolutely certain about where he or she will be.

As Chaoming Song was busily estimating each mobile user's entropy, we already knew that power laws govern our mobility. That is, most people move little, and a few outliers regularly cover hundreds of miles. It was not unreasonable to expect, therefore, that there would be significant differences between the degrees to which our actions could be foreseen. We hoped that it would be easy to make a call on individuals whose lives were confined to a small neighborhood. About people like Hasan, however, who regularly travel over thousands of miles, we knew it would be much more difficult.

This time, however, our intuition failed us—predictability did not follow the familiar power laws. That is, no matter how hard we searched, there were no outliers in the database. Instead, we found, on average, a

93 percent predictability across all users. This means that only 7 percent of the time a person's whereabouts were a mystery. Much of this uncertainty corresponded to transitions between favorite spots—like the unpredictable commute time from work to home during rush hour or variations in lunch plans. For the rest of the day, most users' movements were relatively easy to foresee.

For some users with lower entropy the predictability reached close to 100 percent accuracy. That by itself was not surprising—it was only confirming that some of us are indeed incredibly regular. What was unexpected, however, was that virtually nobody with less than an 80 percent predictability existed in our sample. Notwithstanding the distances they covered and the means of transportation they used, everyone was a prisoner of habit, making their whereabouts easy to predict. There were no Harlequins among our mobile-phone users, making us wonder, What about the spontaneous and the whimsical free spirits? Where are *they* hiding?

Before we move on, let me clarify that there is a fundamental difference between *what* we do and how *predictable* we are. When it comes to things we do—like the distances we travel, the number of e-mails we send, or the number of calls we make—we encounter power laws, which means that some individuals are significantly more active than others. They send more messages; they travel farther. This also means that outliers are normal—we *expect* to have a few individuals, like Hasan, who cover hundreds or even thousands of miles on a regular basis.

But when it comes to the predictability of our actions, to our surprise power laws are replaced by Gaussians. This means that whether you limit your life to a two-mile neighborhood or drive dozens of miles each day, take a fast train to work or even commute via airplane, you are just as predictable as everyone else. And once Gaussians dominate the problem, outliers are forbidden, just as bursts are never found in Poisson's dice-driven universe. Or two-mile-tall folks ambling down the street are unheard of. Despite the many differences between us, when it came to our whereabouts we are all equally predictable, and the unforgiving law of statistics forbids the existence of individuals who somehow buck this trend.

But let statistics forbid, halt, hinder, impede, refuse, and deny, there's still someone who won't be limited by it. Our friend Hasan Elahi.

~

Five years after his Detroit detention and more than a year into his Tracking Transience project, Hasan was flying back to the United States once again, this time arriving at Kennedy Airport in New York on Iberian Airlines flight 6251. Lately, he'd had it easy, mostly free from the harassment of immigration officials as he crisscrossed the globe. But this time, upon disembarking, it became déjà vu all over again—he was separated from the other passengers, taken to a special room, and asked to wait there.

"And you wait, and you wait, and you wait," recalled Hasan. "Someone comes in and asks you a question and leaves and then five minutes later comes back, asks another question, and leaves. You really talk to a messenger; you are never in face-to-face contact with a person that has any authority to take any action."

Eventually, he heard one agent shout to another across the room, "Hey, this guy from that Iberia flight, is he still here?" Much to Hasan's chagrin, he was, of course. And so he was escorted to an interrogation room where he finally learned what the fuss was about.

Not that anybody was willing to tell him anything. This tight-lipped routine—ask questions, reveal nothing—was by now quite familiar to him. But Homeland Security is "a very sloppy organization," as Hasan puts it, so while he wasn't actually supposed to hear or see anything, he couldn't help it.

For example, a piece of paper had been left in the interrogation room, on the table right in front of him. On it was a list of individuals who were to be denied entry to the United States on that particular day, and the list had most certainly not been printed out for Hasan's benefit. Somebody must have inadvertently left it there, revealing to Hasan the names of three suspected terrorists.

They were all Muslim names. One was from Pakistan. The other from Saudi Arabia. The third was from the United States. It was him, Hasan Elahi.

And so he finally understood the reason for his detention. The Pakistani had been detained due to an "active Secret Service case." The man from Saudi Arabia was a suspected arms smuggler. And Hasan—well, he was baffled when he read the cause for detention next to his name.

There was no mention of his ammo-filled storage unit.

Nor was the Tracking Transience project mentioned, his ongoing mocking of the surveillance enterprise.

Instead, the paper said, "Suspicious movement after 9/11."

"Now, just exactly what constitutes *suspicious movement*?" Hasan asked rhetorically, recalling the document. "Not to go off on a tangent, but Clive Stafford Smith, one of the lawyers at Guantánamo, was talking about how they found a report on one of the guys explaining why he was in Guantánamo. And on his file it said that *he exited a taxi in a suspicious manner.*"

Hasan's voice rose in pitch as he continued. "How many possible ways can you exit a taxi? And what constitutes a *suspicious* way of getting out of a taxi versus a *nonsuspicious* way? Or for that matter, what constitutes *suspicious* movement? And what constitutes *nonsuspicious* movement?"

Hasan had no answers. But I couldn't gloss over the issue, as it raised an important possibility: Could it be that *suspicious* means *unpredictable*?

Hasan was not in our mobile-phone database, and we wouldn't have known even if he had been, as the data was anonymized. But since he had been meticulously tracking his own movements for years now, we didn't need his phone records. He sent us a file detailing his every move from February 2007 to December 2007, a ten-month period during which he visited 1,040 different locations throughout the United States and Europe. This may sound like a lot of moving around, but it actually isn't; according to my huge GPS wristwatch, in a two-month stretch in 2007 I, too, visited 515 different locations.*

In some respects, despite our similar mobility, Hasan and I couldn't be more different. Indeed, while I am not as regular in my habits as are the MIT students, at 80 percent predictability my future whereabouts are still quite easy to discern. When Zehui Qu ran the predictive algorithm on Hasan's data, though, it was an epic failure—only three times out of

---

* Zehui Qu broke the map into two-kilometer-by-two-kilometer squares and defined each square as a separate location.

more than four thousand hourly attempts did he succeed in predicting Hasan's movements. We would probably have had more success throwing a dart at a world map. For all practical purposes, Hasan was a Harlequin, fully unpredictable.

I confronted Hasan with our conclusions, telling him that, as far as we were concerned, he was completely random. His predictability was practically zero.

"Can't be zero, is it?" He laughed, then continued without missing a beat, "I mean, there are a few places I go now and then."

Sure, Hasan did visit the same spot in New Jersey 131 times, which, as we later learned, was his home at the time. Even so, he was not easily predictable. By way of comparison, during the two-month period I dutifully toted my GPS device around, I was tracked at home on more than 880 occasions. The difference between Hasan and me boils down to this: While I was predictably at home every night, Hasan was just as likely to be on a train in Europe or asleep in an airport as spending the night in his own bed. He did go home occasionally, but there was no recognizable pattern to it.

From Hasan's perspective, his unpredictability wasn't all that surprising; and while he never explicitly said so, I think he found our whole analysis somewhat puzzling.* That he could readily explain each move he'd made convinced him that his behavior was absolutely normal.

"This is what I do," he said. "It's the transit points that's become my work."

Well, that didn't really cut it for me. Not because I doubted what he said. The real problem is that if power laws had governed our predictability, as they did the distances we cover, we'd expect to have a few outliers. Once power laws are absent, however, outliers are no longer normal. They are forbidden, making us all equally predictable. But no matter how we parsed the data, when it came to his predictability and lifestyle, Hasan was an outlier. Since outliers could not exist in this context, he was not normal any longer. Just as Homeland Security has suspected.

---

* "As of mid-April and on, I was on sabbatical leave from Rutgers," he told me, adding, "so I didn't exactly have a regular weekday schedule to go on. Even when I was at school, I would just basically fly in, teach my class, and then leave. So it does make sense." He then thought a moment, adding, "Because I literally was all over the place that year."

~

Getting back to the young, fabulous, and, usually, fashionable. Are they more spontaneous as a group? Should we attribute the differences between Hasan's and my mobility to our five-year age gap? Not really, we soon learned.

We had access to the records of people between fourteen and eighty-nine years, allowing us to easily compare the predictability of individuals of different ages. The results were unambiguous: Youngster, middle-aged, or mature, everybody's predictability was roughly the same. Only one trend stood out: Independent of age, men were less predictable than women.

We tend to exaggerate the differences among us, believing that we are much less or far more regular than our friends or neighbors. The truth is, however, that you might be a cool *artista,* a rocker, an accountant, or a CEO of a major corporation, but when it comes to quantifying your daily movements, your predictability is likely to be only a few percentage points different from that of the guy next door. As far as our predictability is concerned, we are back to the familiar world of Poisson and Gauss, in which everyone is similar, everything is "normal." You may cover hundreds of miles each day while I cover only three, but you are just as much a prisoner of your habits as I am of mine. The never-ceasing thrill of spontaneity is a mirage at best. Instead, a deeply comforting regularity drives our whereabouts, to a much higher degree than we are willing to acknowledge.

We carry mobile phones, use credit cards, and frequently walk by security cameras, leaving electronic fingerprints all along our path. Harlequins have understood the dangers posed by this electronic trail of breadcrumbs, turning to their RNGs to obliterate habitual behavior. But outside of the world of fiction, I have yet to encounter anyone who would consult an RNG before her next move: Where should we meet later today—get a coffee at Starbucks or fly to Tokyo instead? Let the dice decide. The privacy concerns raised by the proliferation of tracking devices, paired with our deeply rooted regularity, does give me pause, however. Hmm. Shall we fetch the dice?

❧

Despite the high predictability it represents, my low entropy is not a lock on my future—from it you can predict me only if you also know my history. Furthermore, if my entropy is high, my past will reveal little of what lies ahead. If my entropy is low, my movements should be easy to foresee, but only if you have access to my past whereabouts. There is a simple lesson in this, which borders on the banal: In order to predict the future, you first need to know the past.

Uncovering the past, however, is not as easy as it sounds. Take, for example, György Dózsa Székely and his career—how sure are we that the pivotal events we recalled earlier did take place and that they happened exactly as we described them? We are quite confident about some events while entropy veils the rest.

Let me explain. György Székely's duel at Belgrade is so widely noted by contemporaries that its occurrence is rarely disputed. Yet what exactly happened to him and his men between Belgrade and Nagylak is often debated. While it is certain that the cardinal used the pulpit to launch the Crusades, did he really put György in charge there and then? Many historians believe today that in Buda György may have been only one of several officers appointed, by no means a commander-in-chief. Indeed, had the cardinal actually officially named the leader back in Buda, such a momentous decision undoubtedly would have left us a paper trail. No such contemporary proof survives—we know of the cardinal's selection only from chronicles written years after the events.

Given the conspicuous lack of evidence, lately historians have concluded that only after the Crusades turned into an uprising did György Dózsa Székely first surface as its supreme general. If this is true, he may have never planned to march his army against the Ottomans. His Crusade may have had only one target all along: the aristocrats.

Why would the chroniclers insist, then, that it was the cardinal who personally put György Székely in charge? Well, years after the events, after the details of the tumultuous weeks were long forgotten, some people may have found it convenient to believe that György's authority was divine, coming directly from the pope.

The feud between György and Bishop Csáky over the gold the king had promised is another story widely echoed by medieval chroniclers. But perhaps it was nothing more than a benign invention to justify the execution of the unpopular bishop. Therefore, though it may be hard to believe, key events taken for granted by generations of historians and their readers might not ever have actually taken place. Given how impenetrable our past has become, perhaps it's no small wonder that our future is uncertain.

We don't have to search all the way back to the sixteenth century to find an event shrouded in a significant degree of uncertainty. Remember, while low entropy does mean predictability, to predict your future location we need access to your past whereabouts. And as penetrating as they are, phone records are insufficient to the purpose. To predict your future location, I need to know your hourly whereabouts for the past several months. Unless you use your cell phone that frequently—and very few people do—most of the time your location will remain a mystery.

So at the end of the day, whether we are examining the events of today or the sixteenth century, the problem we face is quite similar: If we are not aware of the past, the future is hard to foresee. And what if our past suddenly becomes transparent? Our future, both as individuals and as a society, may cease to be so mysterious. So, in order to reach into the future, we must first go back in time.

And that is exactly what we will do next, visiting the sixteenth century once again to check in on György's efforts to divert attention from his Temesvár siege.

22

A Diversion in Transylvania

Kolozsvár, Transylvania, early July, 1514, midsummer

n 1514 Kolozsvár was not yet the capital of Transylvania, but with its surrounding fertile lands, prosperous guilds, busy market, smelter, and mint it was well on its way to becoming the wealthiest settlement in the country. Once a Roman colony, Kolozsvár had been settled by Hungarians around 895, only to have their descendants massacred by invading Tatars in 1241. Stephen V, the king of Hungary back then, had settled the deserted village with Saxons, but the town's increasing prosperity soon brought the Hungarians back. By the mid-fifteenth century half of the town's four thousand inhabitants were Hungarian, so to keep the peace in 1458 the voivode introduced a shared governance: the judge, the town's effective leader, rotated yearly between the Hungarians and the Saxons.

The generous trade and tax benefits the town received as a royal borough enabled it to modernize its defenses. So a 1.4-mile-long wall with eighteen towers, each maintained by one of the town's powerful guilds, was raised around the city. Drawbridges arched across the surrounding moat and heavy portcullises protected the gates, the weakest spots of any fortress.

But all the peace and prosperity the town enjoyed was about to vanish in 1514. It was just after the summer solstice, the days long and warm, and yet the portcullises were shut, the drawbridges raised, and guild members armed with crossbows and matchlocks crowded the walls and the towers. Meanwhile, in city hall, the town's concerned judge was deliberating with the rest of the council. A large crusading army was filling up the expanse of surrounding fields, demanding entry to the fortress. The town's future and their own lives depended on the council's response.

Had the council members been confident of the town's defenses, there would have been no question as to whether or not they would have unlocked the gates. They knew, however, that despite the well-trained civil guard, they could not hold the walls against an attack by Brother Lőrincz's crusaders. Lőrincz's fame had preceded him, his might and determination proved terrific in his taking several of the better-protected fortresses in Hungary. Furthermore, the town's poor supported the crusaders—or revolutionists, depending on your perspective. So if it came to a siege, the defenders would probably face a rebellion inside the walls as well.

Opening the gates without a fight, however, was out of question. It would not only mean abandoning the wealthy council members and the rest of the citizens to the mercy of the peasants, but would be viewed by the king and the voivode as complicity with the crusaders.

Still vivid in the city's collective memory was King Matthias's return to Kolozsvár, his hometown, after he had quashed a Transylvanian rebellion in 1467. With a sweep of his pen he had revoked that town's status as a royal borough for its citizens' role in the revolt and forced the town to witness the torturous deaths of three noblemen, guilty of siding with the rebels. They had been ripped to pieces with red-hot pincers.

It had taken years for Kolozsvár to regain its trade rights. So once again faced with the horrible army swelling outside their gates, the wealthy councillors of Kolozsvár feared just as much for their lives as they dreaded losing the monarch's favor. Their options were limited, however, as the whereabouts of their protector, the Transylvanian voivode, were unknown. They would be forced to face Lőrincz on their own. But where *was* the voivode, and what was he planning to do about this bursting revolution?

～

In 1505 the national assembly had decreed that if King Ulászló died without an heir, only a Hungarian could be elected to the throne. This was unusual, since in those days kings and queens were chosen from powerful royal families based on their strength and their historical claims to the throne. Nationality had not been a pressing requirement—the concept of a nation as we think of it today barely existed at that time.

The then-king Ulászló was son of a Polish king and a princess of Hungarian origins, and no one saw his simultaneous rule over Bohemia and Hungary as being in conflict.* On the contrary, his double crown had been interpreted as a sign of strength, the alliance furthering both Czech and Hungarian interests.

So when the national assembly called for a Hungarian king, they did so only because they wanted to see the then-eighteen-year-old János Szapolyai seated on the throne. The birth of King Ulászló's son a year later not only ignited a bloody quarrel between the Szeklers and the king's ox-hungry men but also wiped out Szapolyai's chances of ascending to the throne. Yet the young count, richer than the king and leader of the region's most powerful army, did not meekly disappear into the shadows. Under the nobility's pressure, King Ulászló had soon been forced to make Szapolyai the Transylvanian voivode, a powerful office providing Szapolyai with an excellent perch from which to continue his campaign for the throne.

In 1514, after the nobility's forces had fallen one by one to György Dózsa Székely's hosts, it became increasingly clear that only one man had the army to confront the crusaders: the Transylvanian voivode himself, János Szapolyai. So his absence from the action was all the stranger. While he was fighting his own Crusade at the Ottoman border, he could afford to sit on the sidelines, some speculated—true, Hun-

---

* Those kingdoms today would roughly comprise Hungary, the Czech Republic, Slovakia, and part of Serbia and Slovenia, together with Transylvania, today's Romania.

gary burned, but there was peace in Transylvania. There had been no recruitment there, no Crusade camps, and no bloody battles.*

Furthermore, Hungary's misery was not necessarily bad news for Szapolyai. He had opposed the Crusade at its inception, and now he watched with satisfaction as with each passing day more and more of his political opponents were brought to their knees in the chaos. Báthory, his long-term rival, was beaten and now under siege, burdened by heavy debt. Most of the land around Csanád and Temesvár, now controlled by the crusaders, was the property of György Branderburg. The margrave to whom our troubadour Taurinus dedicated his historical poem was a vocal supporter of the Habsburg succession to the throne, thus unlikely to elicit Szapolyai's pity.

But among the camps, the voivode's absence was inexplicable and fueled rumors that spread like the plague among the men. The conventional wisdom was that their captain, György, and the voivode had made a pact to not fight each other. Indeed, given the voivode's long-standing support for the lesser nobility, many of whom now sided with György and his army, they had reason to be convinced that he was on their side.

But when the crusaders marched to the borders of Transylvania, the status quo was disrupted. Brother Lőrincz and his host first reached Nagyvárad, a major fortress and the gateway to Transylvania. The fortifications had been rebuilt and strengthened with enormous walls following the 1474 Ottoman siege, and as an episcopal seat it had the protection of a permanent guard. Not surprisingly, when this prominent town, renowned for its strength and impenetrability, fell to Lőrincz's men, the news reverberated as far as Italy and Prague.

In keeping with his reputation as a fearless leader, Lőrincz had exe-

---

* To be sure, Szapolyai did agree to have the papal bull proclaimed throughout Transylvania, if only to please the court. But the job of mobilizing the peasants had fallen to the bishop of Gyulafehérvár, more interested in arts and sciences than war or politics, who ended up surrounding himself not with soldiers but with notable humanist intellectuals. There was, for example, the prebendary István Stierochsel, who, under the name Taurinus four years later, was to pen in verse the chronicles of 1514; or the archdeacon János Barlabási, Lénárd's cousin and the future bishop of Csanád. Eager to maintain the voivode's trust, the bishop had left for an extended vacation to the faraway Alföld, conveniently forgetting to send the papal bull to his priests before his journey. He made sure to return to his office only after the Crusades had been officially terminated by the cardinal, enjoying the political benefits of his negligence, which spared Transylvania from Hungary's fate.

cuted the fortress's defenders and jailed the nobility captured. But he did not get cozy in Nagyvárad. He left a small force behind, recruited thousands more from the local population, and marched into Transylvania, moving toward the *kincses* Kolozsvár.*

Szapolyai had already crossed the southern Carpathians on his way back to Transylvania when news of Lőrincz's march toward Kolozsvár arrived. So on June 7, in the fortress of Déva, he hurriedly wrote a letter calling for a national assembly of the Transylvanian nobility. The next day, realizing the urgency of the situation, he altered his order. He now demanded that the nobility gather in full armament at Nagyenyed no later than June 25. Each knight must also arm and bring with him one-tenth of his peasants, an unusual request indicating that the voivode recognized the gravity of the situation.

Szapolyai arrived at Nagyenyed a week before his own deadline and from there clarified his stand: Anybody who called himself a crusader or wanted to become one was to be arrested and "decapitated, flayed, or burned and killed, tormented, and exterminated with the most horrible torture."

Aware that he could not count on Szapolyai for help and that he would be unable to hold the city himself against the crusaders, Kolozsvár's judge managed to broker a clever agreement. Giving in to Lőrincz's demands, he opened the gates, but only the officers and their entourage were allowed entry. The remaining army was required to make camp outside the city walls in the surrounding fields. The compromise suited Lőrincz as well, sparing him a lengthy siege.

In return for the city council's cooperation, there was no looting inside the walls and the judge gained custody of several noblemen the crusaders had captured and dragged to Kolozsvár with them. This was his down payment on the future; for by saving the life of several prominent men, the judge hoped the city would redeem itself in the eyes of the king and the voivode for its surrender to the revolutionary forces.

---

* *Kincses* (or *treasure*) Kolozsvár has later become the city's fitting nickname.

He had no way of knowing that Kolozsvár's fall also nicely suited György while significantly complicating Szapolyai's plans.

As ruler of Transylvania, it was now Szapolyai's duty to free the treasured city. He was keenly aware, however, that while such a move might restore his control over Transylvania, it could not win the war. Given his aspirations to the Hungarian throne, now that he was back in the country he could not afford to appear unmoved by Hungary's plight. Furthermore, only a definitive victory was worthy of a future monarch. And such a victory would require that he concede Transylvania to Brother Lőrincz and head instead to Temesvár to confront György Dózsa Székely.

By the time the voivode had arrived in Nagyenyed, he had formulated a plan. He ordered the Saxons of Kronstadt not to send their troops to Nagyenyed, where the nobility was assembling, but to place them instead under the command of the Szekler leader András Lázár of Szárhegy.* This was the self-same Lázár to whom seven years earlier Lénárd Barlabási had referred György Dózsa's fate. In a traditional display of unity between the three nations, the joint Szekler and Saxon units, aided by Romanians from the surrounding villages, were now sent to free the Hungarians and Saxons trapped inside the Kolozsvár fortress.

Szapolyai did not lead this liberating army. He instead asked Lénárd Barlabási, the vice voivode, to take command. The young voivode, his eyes set on bigger quarry, had decided to ride at the front of the columns of Transylvanian nobility and lead them to Temesvár to face György Dózsa Székely.

This strategy had its risks, of course, as it forced Szapolyai to split his forces. With this dangerous gamble he hazarded falling into the same trap that had claimed Báthory, Bebek, Csáky, and many others who time and again had underestimated the crusaders' determination and strength. But it was a chance that the twenty-seven-year-old voivode—yet to be tested by failure—was willing to take.

---

* Today the city of Kronstadt is known as Braşov in Romania and Brassó among Hungarians.

23

The Truth about LifeLinear

In its beta status, LifeLinear's Web portal carried no icon, brand name, or logo. Only an elegant white search box radiated on a black background, which brought to mind a subtle combination of Google's uncluttered interface and AC/DC's sleek *Back in Black* album cover. I typed my last name into the search box, hit ENTER, and watched a short list pop up on the screen. There were only two names:

Albert-László Barabási, Brookline, MA
Dániel Levente Barabási, Wappingers Falls, NY

I clicked on my name and was presented with a familiar portrait, the one of me in my blue shirt, which was accompanied by some basic biographical information, a recognizable Wikipedia rip-off. The rest of the page was filled with dates and links to addresses corresponding to several resonantly familiar locations.

The link I clicked on had a recent time stamp accompanied by a Massachusetts Avenue address in Boston. A video appeared on the screen showing a hurried crowd quickly swallowed by the black doors of the Hynes metro station. About two seconds into the movie I noticed myself, pushing open the station's heavy door. Then I took a left on Massachusetts Avenue and, oblivious to the camera, walked out of its view.

The next link, time-stamped only ten seconds later, returned a photograph of five grinning youngsters posing on the bridge above the Massachusetts Turnpike. At first I recognized no one. But I soon realized that the kids were not the reason the image was placed on my site—the slightly blurred individual in the background was me, captured serendipitously soon after I left the metro station.

I next clicked on a video, a still shot of a short segment of Massachusetts Avenue. This time I entered the camera's view from the left, walked by the Berklee School of Music, and disappeared from view moments later, just before the imposing Christian Science world headquarters.

Watching videos of myself usually makes me squirm. Now, however, I was mesmerized by LifeLinear's offering. So I kept clicking, following myself all the way to my office at the Center for Complex Network Research at Northeastern University. Stupefied, I chose some other dates as well, only to realize that no matter where I was, LifeLinear had a visual on me. Through an amalgamation of video, pictures taken by strangers, friends, or acquaintances, and Web site and blog links, much of my life spent outside the privacy of my home had been cataloged in their vast database.

Identifying an individual in a crowd of thousands from a picture taken years earlier is a challenging computational problem. The task confounds humans as well: a long white beard sufficiently obscured the face of war criminal Radovan Karadžić, arguably Serbia's most notorious citizen, allowing him to live openly in Belgrade for years, unrecognized by the hundreds who interacted with him daily. Considering this, the task of identifying everyone who appears on the billions of images hitting LifeLinear's servers sounds simply impossible. True, LifeLinear's success is not based on a revolutionary face-recognition algorithm; the technology they use is not significantly better than what is available to every other surveillance enterprise. But they nevertheless are able to track a huge fraction of the U.S. population thanks to their one obsession: They have never lost anyone in the first place.

Originally a corporate-surveillance company, LifeLinear installed millions of wireless cameras all over the United States. They combined their feeds into a single searchable database and instructed their computers to keep track of everything that moves. Their technology is based on two principles. The first is known by LifeLinear's programmers as a *conservation law:* Nobody ever vanishes or appears from nowhere. In other words, if you enter a building, a train, or an airplane, sooner or later you will have to exit. We have already encountered their second principle: Our deeply rooted regularity translates into predictability. So LifeLinear builds a behavioral model for everyone, meticulously learning where we normally are and predicting where we will be.

Their system learned, for example, that I typically leave my apartment between noon and 1:00 P.M. Therefore, when their street camera captured my image one day at 12:30 P.M. in front of my home, there was no need to run my picture against the countless images of three hundred million Americans—their software already knew that I was the most likely person to be *there* in *that moment.*

Once I boarded the downtown train, LifeLinear did not even bother to look for me for another twenty minutes. Only around the time my train reached the Hynes station did the scans start on their feeds for my image, aware that this was my most likely destination. As I walked toward my office, the algorithm passed my tag from camera to camera, carefully placing onto my LifeLinear page each segment in which I appeared. Occasionally they fail to locate me at Hynes, which is no great tragedy—their algorithms know that the Longwood station is my second most likely stop, where I get off when I have appointments at Harvard Medical School. Only occasionally, when I break my daily pattern and I get in a cab or board an airplane, must they invest real time resources to track me.

It has probably not escaped your attention that LifeLinear, in many ways, is a reincarnation of the Vast Machine we encountered earlier. It is also the public version of Admiral Poindexter's infamous Total Information Awareness (TIA) program, built to sift through commercial, transportation, financial, communication, and law-enforcement databases under the pretext of the war on terror. But LifeLinear is fundamentally different from TIA and Vast Machines: Those two—TIA in actuality,

Vast Machines fictionally—were designed to comb private and government databases, including banking, e-mail, phone, and FBI records. LifeLinear, however, relies only on data readily available to anybody, like my image caught on security cameras when I walk down the street and personal information that has been culled from the Web. To be sure, all three programs violate our sense of privacy, and many folks assume such programs would automatically be illegal. Not necessarily. U.S. courts have consistently ruled that in public spaces, like a park or the street, we have no reasonable expectation for privacy. Therefore LifeLinear's creators believe that they stand on solid legal ground.

In the final analysis, to most of us the differences between LifeLinear, Vast Machine, and TIA are only cosmetic, and the questions they raise are equally unsettling: Is it indeed possible to track any one of us in real time whether we consent or not? Who would dare to operate such a program? Are we poised to lose our privacy in a world with a fully functional LifeLinear, TIA, or Vast Machine?

Then again, why do we have an expectation of privacy in the first place?

❧

In the small villages of Transylvania, two to three days before any wedding a group of old women swarm the house of the bride to inventory the *perne*—the dowry. It is a business transaction–turned–ceremony, the delicate exchanges between the *perne* women and the bride's family being governed by long-settled customs and habits. Once done with the inventory, a *perne* woman guards the valuables overnight until three horse-drawn carts arrive the next day. The first is for linens, the second for furniture, and the third for everything else.

Warm, hearty *húsleves* simmer in large pots, spicy links of *száraz kolbász* and *szalonna* are laid out on platters, and plenty of glasses filled with *pálinka* keep everybody warm and happy. The father of the bride ceremonially thanks those loading the dowry into the carts, never forgetting to admonish them jokingly to "make sure you don't bring it back!" The litany of puns, proverbs, and poems that follow sound spontaneous to the uninitiated but in fact follow a rigid ritual that is carefully acted out each time a bride approaches her wedding day.

While the family and friends eat, drink, and carry the valuables, one woman carefully supervises the placement of each item. It is her responsibility to find the right spot on the cart for every sheet, towel, bedcover, pillow, and even the bride's childhood doll. The woman isn't preoccupied with making sure everything fits—no, her role is to position the belongings so that each item is *clearly visible*.

Once the packing is complete, the *perne* women and their husbands, dressed in colorful traditional garb, call out joyful verses as they parade the carts through the village. Loud enough to wake the dead, their clear voices prompt women to leave the kitchen and men to abandon the animals, to scrutinize the procession instead. Noisy kids and stray dogs follow the proud cortege, knowing that where there is such a joyful clamor there is usually good food as well. It is a spectacular, long-standing ceremony, with one purpose only: Everybody in the village must inspect the dowry.

The colorful rituals of birth, courtship, marriage, and death in Szekler Land are rooted in the belief that events of such magnitude cannot be private. In fact, they gain validity only if they are witnessed by the community. Those whose birth is not accompanied by a proper baptismal ceremony are said to have been "given a name like a dog"; those who forgo the courtship and wedding rituals are said to be "living together like dogs"; and those laid to rest without the proper sacraments are "buried like a dog." *Isten Dicsőségére, emberek tetszésére,* prays the Szekler, aware that his actions must equally please God and his fellow villagers.* Everything takes place within earshot of everybody else there where the habits and the ceremonies have not changed much since György Dózsa Székely departed on his life-changing adventure. Each villager's life is closely scrutinized so that nothing—love, impotence, theft, disease, hardship, friendship, or hatred—stays hidden for long. Knowing everything about your neighbor is not something to be ashamed of; it is a responsibility, an essential ingredient to maintaining the community's

---

* "For God's glory and people's pleasure."

integrity and well-being. In such a small Transylvanian village, the heightened, pervasive privacy we desire in the United States is virtually unheard of.

None of this is by choice, of course, but by necessity. The inhabitants of this hard-to-cultivate land, frozen by devastating winters and squeezed by the Carpathians' magnificent pine forests, continue to base their existence on a deeply reciprocal work and economic system. To survive they must help one another in both times of peace and times of crisis. Those failing to connect to this network of favors and aid jeopardize their chances at thriving and even surviving, soon finding themselves on the community's fringe.

Watching the Szeklers offers a glimpse of the basic equation governing our privacy: The more a community is interdependent, the less it desires privacy. The more we need our family and friends, the less we can afford the luxury of being tight-lipped. It is only in North America and Western Europe, where our dependence has been monetarized, that we can afford to be left alone. Today research increasingly shows that the key to our happiness and well-being rests in the number and quality of our friends. So, who's to say that *we* got it right? Have we traded happiness for privacy?

⌖

On January 8, 2008, the day I sent the first draft of this chapter to Enikő Jankó, a friend who has been keying my handwritten corrections into the files, I received a text message from her husband, Boldizsár:

"What is the URL for LifeLinear?" he asked, adding, "It appears that Google does not have it."

Amused, I responded: "Why is this important?"

"So that I can see how we partied at my b-day. Why, is this secret?" he wrote back four minutes later.

Happy for the opportunity to tease him a bit longer, I replied, "Yes," only to receive another message: "Then tell me what it is!"

Given my initial refusal, he grew suspicious and soon called to say that his best bet was that I had gone nuts, refusing this information to a trusted friend. His second guess? That LifeLinear does not exist.

He may be right on the first account, but that is not our concern at present. One thing is undeniable: LifeLinear so far is only a product of my imagination. The very fact, however, that Boldizsár did not immediately dismiss it as science fiction shows that it is not so far-fetched.

Let me settle this from the outset: I do not doubt that it is technically possible to develop a system with LifeLinear's capabilities. I also believe that at some point during my life I will live in a world in which some hybrid of TIA, LifeLinear, and Vast Machine will be employed. That does not mean that I advocate or condone building such a system of surveillance. On the contrary, with all the research my team and I have done on human dynamics, with everything we've seen, I get chills whenever I think of such a system's potential capabilities. All I am saying is that science and technology have converged to the point where something like LifeLinear is a possibility. If past technologies are any indication, our misgivings notwithstanding, soon some of its benefits will seem sufficiently enticing for us to accept it.

A Vast Machine or a TIA-like program must continuously feed on data all around us, just as weather prediction relies on current and past atmospheric conditions. And today that data is pervasive: Our time-resolved communications and whereabouts are already available to our mobile-phone carriers; our spending and travel habits are no secret to our banks; our social links and personal interests are documented by our e-mail providers; surveillance cameras regularly tape our behaviors and companions.

Despite the ubiquity of these records, we continue to maintain both an illusion and an expectation of privacy. We hide behind *practical obscurity*, a belief that the many bits of information collected about us are so scattered in disparate proprietary databases that the impediments to accessing and linking them to one another are insurmountable.

The truth is, however, that ever since 9/11 intelligence agencies all over the world have been investing billions of dollars in an effort to stitch together the myriad electronic records routinely collected on each of us. Hasan Elahi's detention for "suspicious movement" after several harassment-free years of travel is a testament to the Department of Homeland Security's efforts to merge private and government databases. While these systems may not yet have achieved the predictive capability of the hypothetical LifeLinear and Vast Machine, they certainly are designed in pursuit of that ultimate goal. And one day they will achieve it unless a decision is explicitly made not to allow it. The question is no longer whether we *could* build a fully predictive LifeLinear but rather who would *dare* to do it. Will it be the government or the private sector?

⁊

In the United States we regularly share our private information with companies in exchange for some real or perceived benefit, like discounts on products and services. We cry wolf in unison, however, if ever we think that the government is collecting personal information about us. Europeans have taken an orthogonal path: Laws forbid companies from sharing personal information about their consumers with other companies, yet all communication companies are obliged, under EU law, to keep and share with the government six months' to two years' worth of records of consumer-activity information, including location and communications records.

In the end, among Americans the perception is that *business is good* and *government is bad*. Europeans operate from the opposite premise: *Government is good* and *private industry is all villains*. Is there such a thing as a universal sense of privacy? If there is, who will enforce it? And lacking enforcement, who will build the Vast Machine? In Europe, strict privacy laws cripple the private sector, so a government-sponsored TIA or Vast Machine is the most plausible scenario. But given the various laws and sensitivities built into the American system and culture, if such a pervasive surveillance were to become functional here, it would likely be a LifeLinear-like product developed as a private enterprise. And there is one company that already has the know-how and the resources to turn it into reality. It's called Google. I'm sure you've heard of it.

⁊

Over the past few years there have been several occasions on which I seriously contemplated shutting down my research on human behavior.

Technology has outstripped our ability to use it responsibly, and I could not ignore the possibility that the fruits of our research would soon become part of some malicious Vast Machine–like enterprise.

It took the news coverage following our research paper on human mobility for many people to realize just how much information is already collected on each of us. For some, their first instinct upon reading our findings was to shoot the messenger, comparing us to Big Brother. The sleepless nights that followed forced me to ask myself, what exactly is the role of the researcher? This is not a particularly novel question, having haunted generations of scientists, involved in everything from nuclear energy to genetics. Just like human dynamics, these fields offer tremendous benefits, from novel drugs to clean energy. But they have their dark side as well, from nuclear weapons to Frankenbugs.

Today each person doing research on human dynamics increasingly faces a similar dilemma: How do we avoid contributing to the creation of a surveillance state or conglomerate, a back-to-the-future ticket to Orwell's *1984*?

Hasan has a refreshing answer to this question.

"Intelligence agencies, regardless of who they are, all operate in an industry where their commodity is information," he observes. "The reason their information has value," he adds, "is because no one has access to it."

His solution? Give it up, and it becomes worthless. "It is the secrecy applied to the information that makes it valuable," he says. And with that, he joins the Szeklers and hides in plain sight, pouring his life out onto his Web site.

But does he really abdicate his privacy? If you visit Tracking Transience, the Web site on which he tracks himself, you will soon notice that Hasan never, not once, appears in any of the tens of thousands of images he posts there. Yes, yes, he's on the other side of the lens. But there are hardly any recurring faces in the catalogs of photos, either, and no matter how long you browse his images, you can't escape the sense that he has no colleagues, no family, no friends. The more you explore his site, the more you are forced to ask, What exactly is all this? Why am I looking at this guy's toilets and meals?

"In a sense, I'm giving you everything and I'm giving you nothing," he told me once. "In that noise of data, as open as my life is and as much information I give, I really live an extremely private life. You may know all these financial details about me, and you may know where I live, you may know what my house looks like—you may know everything about me, but you know absolutely nothing about my personal life. So in a sense, it's a very counterintuitive thing, but I protected my privacy by completely giving it up. Because when everything is up front, no one cares."

Unfortunately, when it comes to the inherent tensions between research and privacy, I have yet to determine how to untie the Gordian knot. Giving up academic research would only push human dynamics into secretive government labs and the tight-lipped private sector, free of any oversight. Indeed, today more research is done on human activity in private companies than in universities. The AdSense program, Google's cash cow, is nothing but a huge experiment in human behavior, aimed at tailoring advertising profits to maximize quarterly earnings.

So do we even need academic research? Beyond the thrill of discovery it affords, I feel responsible to use it to bring to the public's attention both the promises and the limitations of emergent technologies. I dreamed up LifeLinear mainly with this purpose in mind, to illustrate the potential outcome of this research. There are countless tools out there, legal and technological, to stop it from becoming a reality if we really don't want to live in a world monitored by it.

~

As we ponder the fate of research on human dynamics, one question, too often ignored for its absurdity, insists on my attention: Who owns our future? Currently there are countless privacy laws, regulations, and practices to protect the data collected on us. Our information is further safeguarded by a healthy dose of fear and caution: From our e-mail providers to our mobile-phone carriers, most companies' business models are too lucrative to risk by mishandling our personal information and angering the consumer. So it is safe to say that despite the many potential risks represented by the volumes of data available, our past is relatively well safeguarded.

But what about our future? How well is that protected?

As we have seen, predicting an individual's behavior is getting steadily easier. And the future is far more valuable than the past, as our travel and purchasing plans are possibly the most potent commodity in our economy. And while secure firewalls and privacy laws protect our pasts, our futures, predicted by sophisticated algorithms, are up for grabs. With that we arrive at a new paradigm I call *prospective privacy*. It boils down to this: Who owns the information about our future actions and behavior? Who should profit from it?

This is a strange shift in perspective, one that we need to apply to our historical drama as well. For while everything we have so far encountered took place in the distant past, we have yet to reveal its outcome, which, of course, happens to still be in the distant past for us. The forces that will draw our conclusion have slowly begun to reveal themselves: After a rapid campaign that brought much of Hungary under his rule, György has settled at Temesvár, where he is attempting to establish a permanent base. Brother Lőrincz, his fearless lieutenant, has forced open the gates to the prosperous Kolozsvár. And so the sleeping lion awakes. Count Szapolyai, the voivode of Transylvania, will not, cannot, remain unengaged now that the uprising has reached his empire. Faced with a war on two fronts, he thus chooses a daring strategy, splitting his forces between the two battlefields.

A cascade of events has been set into motion. It is time for us to watch them unfold.

24

Szekler Against Szekler

July 15, 1514, less than two months after the massacre

ow's your chance to punish your accursed enemy,"* bellowed Gyorgy Dozsa Szekely, forcing his voice to be audible to every soldier in the vast army before him. With a sensation of déjà vu, he scanned the sea of men who looked quite like those he had led into battle at Nagylak two months earlier—merchants, smiths, weavers, tailors, and other townsfolk lost among the swarm of outlaws and peasants. This wasn't Nagylak, however, but the expansive Ulics field next to the Temesvár fortress, and the difference two months had made was evident to the warrior's practiced eye: A great deal more armor was strapped over the dirty tunics, bounty from past victories, and many had ditched their axes and scythes in favor of swords and spears.

The campaign had taken its toll on all of them—this lot was much rougher, more desperate. They were no longer the mindless, stampeding animals that had run blindly into battle and equally quickly fled at the

---

* After Giovanni Michele Bruto (1515–1599), *Historia Hungaria*.

sight of a fast-paced mounted knight heading their way. Battles and hardship had purged their ranks—the weak had fallen, and those whose courage wavered had by now run away. The survivors, shielded by a false sense of invincibility, had morphed into an oddly disciplined army emboldened by the sweet rush of victory to which they had become accustomed.

"The time has come to fight for the freedom of your loved ones," continued György with his usual self-assurance, acutely aware that today of all days he must exhort every bit of vim and vigor his men had in them. For on the gently rolling hills in front of them, wearing colorful surcoats and gray armor, Count Szapolyai's Transylvanian forces were massing. It was a formidable foe, György knew only too well, having drawn his sword under its banner countless times while a mercenary at the Ends.

Szapolyai's front lines were filled with grim Saxon handgunners wearing visored German sallets, each carrying a matchlock and a crossbelt of cartridges with a bullet bag. Behind them massed thousands of confused peasants armed only with sword and spear. Though their numbers were great, they failed to impress the Székely; he knew that his peasant forces were far more experienced and determined than these poor souls dragged to the battlefield under the voivode's threats.

A line of cannons separated the enemy's foot soldiers from the cavalry that comprised the expansive left and right flanks. They were a mixture of heavy knights and light hussars, grouped in *conrois* between twenty and forty strong. Their embroidered tunics and bright banners rendered them the most visible of the opposing force, their agitation evident as they rode up and down the lines, rallying one another for the assault soon to come. It was they that most worried György; he understood that battles were won by fast and mobile cavalries. His crusaders, mostly on foot, were no real match for those mounted Transylvanian knights.

"Take vengeance on those who would ruin your country; fight for a just cause against the most cruel enemy," concluded György. As his men cheered, the Székely's thoughts kept returning to the small but forceful cavalry behind the Transylvanian infantry. They were his blood, the Szeklers; he had recognized them immediately on their short mountain horses. Their placement and numbers had surprised him, however. Traditionally it was the Szeklers' duty to be the first to charge the enemy.

But today the voivode had placed them behind his main forces, together with the mercenaries and his palace guard, apparently keen to shield his most valuable forces from the worst of the battle. Or was he afraid that the Szeklers might switch sides, choosing to obey their own blood rather than their voivode? In the end, it was a question of priorities.

But most puzzling to György, more than the Szeklers' placement, was their number. Where were the rest of them? When attacked, the Szeklers could arm as many as thirty thousand men and easily send a ten-thousand-strong force when called to battle by their voivode. Yet this Szekler contingency was tiny, numbering maybe a thousand altogether. What, György Dózsa Székely wondered, did the voivode have planned for the rest of his tribe?

Despite the obvious strength of Szapolyai's forces, György could spot its weaknesses—not only Szeklers but important Saxon and Romanian units were nowhere to be found either. Their absence had to be the reason why the voivode had augmented his ranks with inexperienced peasants. So the crusaders' superior numbers and resolve gave György hope that even without a potent cavalry he could win this battle as well.

Meanwhile, two hundred miles from Temesvár, in the city of Kolozsvár, the poor had taken matters into their own hands. Encouraged by the crusader captains who had established quarters in the town's taverns, the paupers robbed the nobility who had taken refuge within the walls of the fortress. Things weren't too encouraging outside the walls either—the crusaders had slaughtered the town's cattle and confiscated all the produce the surrounding lands had to offer. The growing crisis unnerved the town's council as they realized that they were not only powerless against Brother Lőrincz and his men surrounding the fortress but had lost control over the events within their walls as well.

With time, however, the mood of the town began to change from jubilant defiance to fearful confusion. The scarcity of food, the authority vacuum, and the receding order and security curbed the poor's enthusiasm for the crusaders. So they simply stood by when the judge firmly changed course and, in a surprise move, lowered the portcullis, cutting the city off from the malevolent throng camped in the fields. Inside the walls he had every crusader captain arrested.

Brother Lőrincz must not have been in the fortress at the critical moment, for we know that he escaped unharmed. And the judge's about-face hadn't been entirely impromptu; he must have gotten word that help was on the way. Indeed, in the chaos that reigned while the crusades' captains were imprisoned, Vice Voivode Barlabási approached the crusaders' camp outside the city walls largely unnoticed, helming a host comprised of those Szeklers, Saxons, and Romanians who had been so conspicuously absent from Szapolyai's lineup opposing György at Temesvár.

And so, by mid-July the battle lines had been drawn. At Temesvár, György Székely's crusaders faced the smaller but more experienced Transylvanian forces, led by the voivode himself. At Kolozsvár, an army of Szeklers, Saxons, and Romanians exploited Brother Lőrincz's crippled command, following the vice voivode in ambush of the crusaders' camp.

Two battlegrounds, each with uncertain outcome. Yet the stakes were clear to everyone: If the crusaders were to prevail against the voivode, there would be no force in all of Hungary or Transylvania left standing that could halt them again.

25

Feeling Sick Is Not a Priority

I n 2006 Nicholas Christakis saw firsthand what it means to let a genie out of his bottle. And he would tell you that it isn't always pretty. Tall and animated, in his mid-forties, Nicholas is a dynamo, shoehorning with his impatient curiosity several apparently divergent research interests into the space of his twenty-four-hour days. With appointments in both medicine and sociology, for many years his research has focused on the widower effect, the observation that elderly spouses tend to die within a conspicuously short time of each other.

One day he started thinking that perhaps the phenomenon was not limited to spouses and death. For example, could changes in my health impact the well-being of my acquaintances? Obviously infectious diseases—like influenza, SARS, or AIDS—do spread from person to person, but could our friends be responsible for our noninfectious diseases as well? Could my heart attack land my best friend in the emergency room with heart failure? Could *his* cancer send *my* cells down a cancerous pathway?

To one extent, the hypothesis was so ridiculous that only asking it could have been career-ending. But as a tenured professor at Harvard since 2001, his job secure, he had only his reputation to lose. And so he teamed up with James Fowler, a political scientist with a keen interest in networks, and the pair headed off to a small town near Boston.

Probably the most famous health survey ever, the Framingham Heart Study was initiated in 1948 when researchers from Boston University began an extensive set of physical examinations and lifestyle interviews of 5,209 men and women from the town of Framingham, Massachusetts. In 1971 they enrolled the participants' children and in 2002 their grand-children, asking them all to return every two years for physical examinations and a battery of laboratory tests. The impact of this monumental study cannot be overestimated—much of our current knowledge about heart disease, from the role of cholesterol to high blood pressure, has its roots in Framingham.

Nicholas and James had little interest in heart disease. They knew, however, that they could save quite a bit of money and resources if they were to pick their research subjects from Framingham, as the disease history of many of the town's residents had already been cataloged. All Nicholas and James needed to do next was identify their subjects' friends and collect their disease histories as well.

Despite their attempts to be frugal, the project still would cost a staggering $25 million. The National Institutes of Health, to whom they had turned for support, was not thrilled. Before so much of the taxpayers' money would be spent on an untested theory, the duo needed to amass some preliminary data, they were told—anything that might indicate that the whole concept was more than a hypothesis. So Nicholas returned to Framingham, with somewhat deflated goals, and began planning a preliminary study.

"Not to be self-congratulatory," he told me, "but I am very particular about my data. I need to know everything about it—how it was collected, who collected it, and what each entry means." And to satisfy his curiosity, he had tried to meet with everybody involved in data collection at the Framingham Heart Study. One day, in so doing, he met with a woman who explained the ropes to him.

She pulled out a green sheet they had used to record the data pertaining to each participant and said, "Here is you, where you live, where you work," pointing to the standard demographic information. She then pointed to the columns recording information about "your brother, your family members, and your best friends."

Nicholas couldn't believe what he had just heard. "At that moment I realized, 'Oh, my God, they already have all the data we need for our study!'"

Collecting information on friends was not part of traditional survey procedure. But given the long duration of the study planned, the doctors in Framingham had been concerned that participants might move away without leaving a forwarding address. They reasoned that even if they did that, their best friends would know how to reach them; and so with only keeping tabs on their patients in mind, the first Framingham researchers began carefully recording each person's social contacts.

And there was yet another stroke of luck for Nicholas and James. Given how small Framingham had been in 1948, most of the participants' friends and their friends' friends also already happened to be part of the databases. So in the end there wasn't any need for the NIH's $25 million to explore whether diseases spread to acquaintances—everything Nicholas and James needed to test their theory was already right there on the green sheets.

To make their first study convincing, they wanted to focus on a condition whose diagnosis did not depend on self-reported symptoms. After some searching they found one that fit their criteria: obesity. To diagnose it, all you needed was the individual's body-mass index—or BMI—which is measured as the person's weight (in kilograms) per the person's height squared (meters squared). The diagnosis left no room for ambiguity: Individuals with BMI above thirty kg/m^2 were considered obese, and those above twenty-five were overweight. Most important, each participant was weighed every two years, offering the researchers a three-decade-long weight history.

The results were beyond comprehension. Nicholas and James found that if a friend of yours were to become obese, the risk that you, too, would gain weight in the next two to four years increased by 57 percent. The risk tripled if it was your best buddy who ballooned—in this case, your chances of following suit jumped by 171 percent. For all practical purposes, obesity was just as contagious as influenza or AIDS, passing along our social links from person to person.

The simplicity of their conclusion was startling: To avoid putting on weight, make sure your friends don't. Yet their findings weren't that easy to swallow. An estimated 65 percent of U.S. adults are already overweight, amounting to a staggering 127 million people, of whom 60 million are obese. Galled by headlines like "FAT FRIENDS ARE BAD FOR YOUR HEALTH," many were insulted by the results—particularly after the press labeled them "obesity carriers" or risk factors for their friends. Soon, Nicholas and James were flooded by phone calls, e-mails, and letters, not all of them congratulatory. Some thought that "these so-called eggheads at the universities must be on drugs," as the results were obvious, while others argued that the research was deeply flawed. "It was hysterical—we really wanted to introduce these people to each other and get out of the way!" recalls Nicholas a little gleefully.

The findings quickly morphed into a cultural phenomenon, the genie starting a life of its own. Jay Leno began using it as a punch line on *The Tonight Show,* and William Shatner's *Boston Legal* character Denny Crane threatened to fire his obese assistant for the health risk she posed to him. Not everybody was laughing, however. The mood of many was well captured by this heartbreaking comment from a reader of *The New York Times:* "The article made me very depressed. I am certain I will have less friends after it was published because I am fat."

～

Several months into my 2005 sabbatical at the Harvard Medical School I had made little progress in answering the question that had brought me to Boston in the first place: Is the timing of our diseases random, or does it follow definable patterns? Can burstiness affect our well-being? I knew exactly what I needed to move forward: the disease history of a large group of individuals. E-mails to my requests had gone unanswered, however, and my hottest leads went cold. My attempt to gain access to the appropriate data sets was floundering.

Then, one day I caught my lucky break. Following a talk I gave at the

Swiss Consulate in Cambridge, a tall, buoyant man approached me, cheerfully asking if we could chat. We did, and a few days later his booming voice filled the small conference room with talk of social networks and obesity. But it was something he had mentioned only in passing that really got my attention. My new acquaintance, as you might have guessed, was Nicholas Christakis, and by that point he had already collected the disease histories of a million elderly couples in his attempt to explain why spouses die together. And this was exactly the kind of data I needed to determine whether bursts affect our health.

Whenever anybody sixty-five years of age or older enters a doctor's office or a hospital in the United States, a detailed record pertaining to the visit, including its time, its location, and the diagnosis, is submitted to Medicare, the government-run health insurance program that covers the cost of the treatment. Therefore, Medicare has the disease histories of most elderly Americans, and Nicholas had access to a ten-year sample of it. So, with his help, Kwang-Il Goh, a postdoctoral researcher in my group, assembled the full sequence of the times that two million patients had seen a doctor, whether they visited their primary-care physicians, specialists, or ER staff.

"We know neither the time, nor the day, nor the hour, nor the minute," the Szeklers say about death, which applies to disease as well. Indeed, when and what diseases we develop depends on a large number of factors, from the genes we have inherited to our diet, exercise, smoking, and drinking habits, to the kind of work we do and our environment. Therefore, the precise day that symptoms to a disease emerge is expected to be random and, hence, unpredictable.

And so this was one of the reasons why we were so surprised to find from our data set that none of the individuals under study had a random-looking disease history. What we saw instead were long intervals during which patients avoided the doctor's office altogether, representing extended healthy stretches. These were followed by short periods during which the patients visited their doctors so frequently that it was as if they'd moved in together. The health histories were bursty, resembling more closely our Web-browsing and e-mail behaviors than a sequence of events driven by a throw of the dice.

Once you accept that none of our activities is random, the burstiness we found in the disease histories is perhaps not quite so surprising. At the time, though, we were taken aback. You see, the emergence
of an illness depends on everything *but* our priority list. If we could
actually prioritize the onset of diseases according to our convenience,
I am certain they would all wind up at the very bottom of our to-do
lists. By assigning them a low priority we could make sure that we
would never fall ill, staying busy and healthy throughout our entire
lives. Sadly, it doesn't work that way—diseases "strike us," making us
sound like victims randomly, unpredictably, inconceivably, felled by a
capricious attacker.

Yet as burstiness greeted us in the records of patient after patient, we
redoubled our efforts to figure out exactly *why*. After all, it's people's
health, their very lives, we were talking about. And finding the answer
required that we take an unexpected turn, immersing ourselves in the
magic of both Harry Potter and the curse of the Bambino.

ॐ

It's during the weekends of summer vacation that kids are finally able to
climb trees and run through their neighborhoods, venting all the pent-
up energy collected over the course of the school year. It's also, not coin-
cidentally, the time when most accidents occur. Thus, June, July, and
August see emergency rooms fill quickly with fractures of the thin and
fragile forearm, wrist, and ankle and with more serious head injuries.
And so it was entirely in spite of the odds that on the weekend of July 16,
2005, all was calm, all was bright, in trauma surgery at John Radcliffe
Hospital in Oxford, England.

Of course, each accident follows a unique sequence of events that
makes its timing rather unpredictable. But even randomness has its inner
order—precisely because each accident is individually capricious, the
load at the emergency room is somewhat predictable, just as the load on
the phone tower is predictable if people make random calls throughout
the day. At John Radcliffe this load averaged out to slightly fewer than
seventy cases each weekend. The previous weekend had been somewhat
busier when almost eighty patients were treated. On July 16 of that year,

things were different—by some miracle, the little patients had managed to stay out of trouble.

The whole thing just didn't make any sense. The weather was good, the schools were closed; and, it followed, many, many more bones should have been broken.

"We were twiddling our thumbs," recalls Stephen Gwilym, a doctor in the emergency room at John Radcliffe. There were so few patients that Gwilym eventually encouraged his colleague, Dr. Keith Willett, to go home; one doctor could easily cover the unit.

As a father of five, Dr. Willett normally would have jumped at the chance to spend more time with his family. On this day, however, he hesitated. There really wasn't much point in going home, he explained, as four of his five children had camped out on the sofa, reading Harry Potter, an all-day venture. They had no use for their father today. Just as Willett said that, the proverbial light bulb flashed above the doctors' heads. Could it be that some Potter magic was responsible for keeping the kids out of trouble?

As any decent Harry Potter fan is surely aware, July 16, 2005, was no ordinary Saturday in Muggleland. That day saw the much-anticipated release of *Harry Potter and the Half-Blood Prince,* the sixth installment in the hugely popular seven-volume children's series. It had sold almost nine million copies on that first day alone, as Dr. Willett knew too well; he alone had bought four of the books to prevent fighting among each of his children of reading age. But was there *really* a link between the *The Half-Blood Prince* and the empty ER?

Curious, the doctors pulled the hospital records for the weekend of June 21–22, 2003, when the previous Harry Potter installment, *The Order of the Phoenix,* had come available. Sure enough, on that weekend, too, there had been an abrupt drop in ER cases—to fewer than half the normal number. These were staggering numbers. At no other point during a three-year period had visitation been so low, prompting the doctors to propose that distraction therapy might be the best way to avoid childhood injury. All we need are "safety-conscious, talented writers who could produce high-quality books for the purpose of injury prevention," the doctors concluded, tongue in cheek.

It is tempting to view Harry Potter's impact on broken bones as an amusing anomaly with a straightforward explanation: When kids read, they stay out of trouble. The phenomenon is not unique to Harry Potter, however. Indeed, researchers at the Children's Hospital in Boston, Massachusetts, were similarly stunned by the impact the local ball club's fortunes had on area patients.

Prior to 1918, the Boston Red Sox had been one of the most successful teams in baseball history, amassing five World Series titles. Then they sold the New York Yankees the Bambino, Babe Ruth, the legendary hitter. With that, the once-lackluster Yankees evolved into the most successful franchise in the Major Leagues, while the Sox bore the so-called curse of the Bambino, failing to win a single championship for the next eighty-five years. This was quite a blow not just to the Olde Towne Team's fans but to the health of the Bostonians in general—hospital records indicate that during losing games emergency-room visits were up by about 15 percent.

Only in 2004 did the Red Sox start to win again. Sure enough, during the final two games of the ALCS championship and the final game of the World Series the Sox won their first title in almost a century. And emergency-room visits dropped by 15 percent.

It is normal to think that our visits to the doctor's office are driven by pain and disease, which is surely random and unpredictable. Yet Harry Potter and the Red Sox suggest that our well-being is more linked to our priorities than we are willing to acknowledge.

The larger issue at play here relates to *when* we decide to seek help. If our symptoms require immediate attention, like a broken wrist or unbearable abdominal pain, *when* means *right away*. For most diseases, however, the initial symptoms are subtle—headache, tiredness, migraine, joint pain—making our doctor visits discretionary. We may wait a day or two to see if the symptoms go away, as they often do. If the Sox are doing well, we may not even notice the discomfort. If they lose, however, seeing a doctor is less painful than watching the Yankees' home runs.

In the end, the less severe our symptoms, the more likely a doctor's visit will sink to the bottom of our priority list. And as we now know, once priorities are in play, burstiness soon follows—we quickly address the top items on our lists while a few tasks (like seeing the doctor) can languish indefinitely. And this is one reason why health insurance makes so much sense: It is practically impossible to time the occurrence of our diseases. If our disease histories were random, we could budget approximately the same amount of money each year toward our health related expenses. Thanks to burstiness, however, there are long periods of health when much of the insurance premium appears wasted. Once one disease does strike, it can unleash a wave of interconnected events, resulting in not one but multiple visits to the doctor's office. But there is a silver lining in all this—burstiness ensures that sooner or later we will enjoy another long healthy stretch.

⁂

By 2020, depression will likely be second to only heart disease as our nation's deadliest killer. It is one of the greatest health scourges of our age, affecting approximately eighteen million Americans, and it is often lethal. Up to 60 percent of all those who commit suicide suffer from mood disorders. Depression is also one of the most misunderstood diseases, as more than half of the U.S. population views the condition as little more than personal weakness. Its stigma is rooted partly in its diagnosis—doctors rely mainly on self-reported symptoms, which are, by definition, subjective. And so if technologies were developed that could codify a depression diagnosis as clearly and stringently as we can diagnose cancer or heart disease, not only would millions be helped, but the confusion and ignorance surrounding the debilitating condition would be wiped out as well.

A common symptom of depression is a visible slowdown of gestures and movements. "You're encased in cement, where you just can't drag your body out of bed," recalls one depression sufferer, making us wonder, does depression change a person's activity, or is this feeling only the impression of an afflicted brain? Since by now we know what constitutes a normal activity pattern, we can ask, do depression patients do anything differently from the rest of us?

It was a Japanese research group from Tokyo University who first addressed this question. They provided twenty-five individuals sensitive wrist accelerometers, designed to pick up on even the slightest movement of the hand. The detectors confirmed that human activity is bursty, down to the slightest movements of our wrists. Indeed, the researchers determined that the length of periods of rest—when the subjects' hands did not move—follows a power law. Most rest periods lasted only seconds, or minutes at most. Yet these brief pauses coexisted with movement-free intervals lasting hours, capturing sleep, rest, or meditation.

Fourteen of the twenty-five subjects displayed a more intermittent rest pattern than did the others. They were clinically depressed. The differences in their movements were striking: The average rest period in the healthy subjects was about seven minutes, compared to over fifteen minutes for depression patients. Furthermore, the scaling exponent, a number that uniquely characterizes each power law, was larger for healthy individuals than for the depressed. And so it would seem that being "encased in cement" was more than a mere illusion, actually corresponding to detectable changes in a depressed patient's activity pattern.

Normally the path from basic science to its application is frustratingly slow. Quantum mechanics, the scientific triumph of the twentieth century, did not offer any tangible benefits for about a half a century, until the transistor's discovery. Similarly, despite the medical revolution sparked by the decoding of the human genome, a decade later virtually all drugs on the market are still the result of the trial-and-error process of pregenomic medicine.

Given these timetables, I was surprised to see how quickly the concept of burstiness has moved from basic science to application. Indeed, you needn't have a Ph.D. to understand the potential implications of these findings. Among other things, they open the door to simple, nonintrusive depression diagnosis. You're feeling low, and the symptoms point to a potential mood disorder? Wear a wristwatch that tracks your movement, and the diagnosis will be waiting for you at your doctor's office, ready to fend off the depression looming on the horizon.

~

The more we learn about the many things that can possibly go wrong in our cells, the more of a miracle it seems that we are even occasionally healthy. If a mutation in the P53 gene handicaps the protein's ability to kill corrupted cells, we will soon develop cancer. If a misfolded protein prompts other proteins to similarly misfold, mad cow disease could follow. If the serotonin concentration in our neurons drops, we are at risk for depression. Then consider that it is harder for two proteins to find each other than it is for you and your best friend to serendipitously bump into each other while wandering New York City, and you start to wonder how our genes ever succeed at their job to the degree they do.

You are not alone. Biologists continue to puzzle over the cell's ability to coordinate the activity of the myriad of genes, proteins, metabolites, and RNA molecules that make up our tissues. We know so little about these processes because it is extremely difficult to peek into our cells.

Enter Ido Golding, a physics Ph.D. who joined the biology lab of Edward Cox at Princeton University and set out to capture the precise moment when individual RNA molecules are produced by a gene. It is no easy task to generate an RNA molecule. First the cell needs to assemble an RNA polymerase from several components. At about the same time, other proteins and metabolites must find the precise spot on the DNA strand where the polymerase will have to attach, an incredibly difficult process in its own right. Given the many unpredictable factors that contribute to RNA-molecule generation, standard theories assume that it occurs at random, unpredictable intervals, following a Poisson process.

Ido Golding's measurements showed anything but a Poisson sequence, however. Instead, he saw a clear *on* and *off* pattern. That is, in a period lasting anywhere from one to fifteen minutes, a single gene was typically turned on, during which time it produced in close succession between two and seven RNA molecules. These bursts were followed by long inactive periods, lasting anywhere from ten minutes to several hours. Cellular activity is not random, but neither does it move with the precision of the second hand on a Swiss watch. Instead, cell movement is an amalgam of the two, and our genes limp along in a haphazard, bursty fashion.

In a completely different era and domain, Charles Darwin hypothesized that the emergence of each new species was a gradual process, taking place through the slow transformation of existing species into their somewhat-modified offspring. Yet evidence for such continuous change was not only lacking back then but is scarce even today, having prompted Darwin to label it "the gravest objection [that] can be urged against my theory." Instead, over millions of years species in the fossil record show little or no evolutionary change. New species tend to appear over periods spanning tens of thousands of years, a split second in terms of all evolutionary time. Evolution proceeds in bursts, which are in turn preserved in the fossil record.

The phenomena we encountered in the previous chapters, from e-mail usage to travel patterns, hint that burstiness is deeply linked to human will and intelligence. Prioritizing only reinforces this impression, since it is our preferences that determine whether an action item is seen to immediately or indefinitely postponed. This would suggest that bursts require the ability to set priorities. But from this perspective, the results discussed above are rather humbling. They indicate that burstiness is not something we invented but was in use well before intelligent life ever emerged on Earth. There's nothing smooth or random in the way life expresses itself, but bursts dominate at all time scales, from milliseconds to hours in our cells; from minutes to weeks in our activity patterns; from weeks to years when it comes to diseases; from millennia to millions of years in evolutionary processes. Bursts are an integral part of the miracle of life, signatures of the continuous struggle for adaptation and survival.

As fascinating as they are, these findings also unleash a flood of perplexing questions. For starters, if it is not only decision making and priorities that produce bursts, why, then, do bursts emerge in so many systems? Can we understand their prevalence?

Recently system biologists have developed gene-activity models that capture the bursts within our cells. In the 1970s evolutionary biologists Niles Eldredge and Stephen Jay Gould presented a new theory they called *punctuated equilibrium,* which describes the rapid evolutionary changes observed in paleontology. By no means did they solve the entire mystery, however. On the contrary, the visible gaps in our knowledge leave room for a deep philosophical conundrum: Are bursts proof of Mother Nature's frugality, converging to similar solutions in very different environments? Or might they represent different facets of some deeper reality?*

Knowledge seems also to evolve in a bursty fashion, a spark turning the spotlight on something that was hidden for centuries. Then again, once we find a solution, have we really resolved anything, or have we actually merely bred ever more questions? These two are not mutually exclusive, though, as most revolutions in the realm of ideas or science are better known for the doors they opened rather than those they closed forever.

In any case, when it comes to history, some doors have been closed for centuries. For example, we can never know with certainty the motivations of the actors in our historical play. Even so, there is little ambiguity regarding the outcome of these past events.

Alea iacta est, said Julius Caesar as he crossed the River Rubicon and thus started the great Roman civil war. Another Rubicon has been crossed in our Crusade-cum-peasant-revolt now that Szapolyai has returned to Transylvania in pursuit of dominance and the Hungarian crown, splitting his forces into two. György faces Szapolyai at Temesvár, and Lőrincz faces Barlabási at Kolozsvár. As our characters spring into bloody action, we lean back and watch them play out their hands. Then we will return to our prophet, Telegdi, and inspect his scorecard, asking finally: How accurate was he and how did he do it?

---

* In the 1980s and 1990s a rather popular and influential theory, called self-organized critical-ity, introduced by the late Per Bak and his collaborators Chao Tang and Kurt Wiesenfeld, aimed to address this very question using a battery of simple models and general principles.

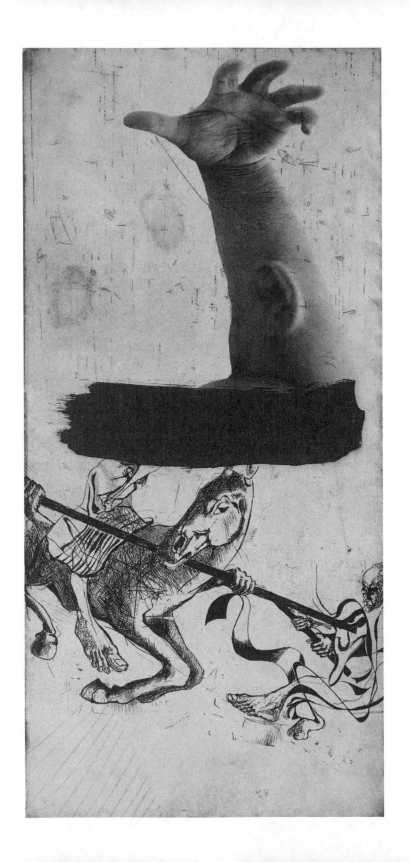

26

The Final Battles

July 15, 1514, less than three months after the Mass that launched the Crusade

ost chronicles agree on one detail: that the two armies at Temesvár did face each other. They tell us that György Dózsa Székely commanded the main forces of the crusading army, entrusting one of the two flanks to his reluctant brother, Gergely, and the other to Father Lőrincz.* János Szapolyai captained his infantry, while two of his lieutenants headed each of the two flanks of mounted men. What happened after that . . . well, that remains a bit of a mystery.

Given the importance of the Temesvár battle, it is surprising how little the various accounts of the events agree with one another. Several of the chroniclers—Brutus, Jovius, Istvánffy, Taurinus, and Bartholinus—describe a bloody battle. Other sources, strangely, flatly deny any bloodshed.

According to the narrative echoed by most history books, Szapolyai reminded his knights of the untold legions of noblemen whose lives had

---

* This is Father Lőrincz of Cegléd, no relation to Brother Lőrincz of Bihar, who led the Kolozsvár crusaders.

been brutally cut short by the axes of the savage peasants. He promised an easy victory over an enemy whose guilty consciences would force them into a cowardly retreat at their first encounter with the superior Transylvanian army.

Szapolyai then sent an envoy to convey his message for the rebels: Surrender, and go home unharmed; place a branch with green leaves on your head as a sign of peace, and the knights will spare your life. Fight, however, and perish, for no mercy will be shown to those who insist on defying the nobility.

The message was familiar to most of the peasants, as similar promises had been made before earlier battles. But they weren't entirely sure if the offers of clemency were sincere, as up until now none of their lords had lived long enough to try their hands at mercy.

This time, however, the battle had been preceded by an exhausting month-long siege of dubious success, and rumors of Crusade losses had been trickling in from all over the kingdom. So some of the crusaders, tired and demoralized, were paying close attention to Szapolyai's offer.

Radoszlav was one of them. The Serbian captain had led the forces responsible for conquering much of southern Hungary before the nobility pushed him back all the way to Temesvár. And now he was ready to accept the noble pardon and lay down his weapons. According to two contemporary sources, when György Dózsa Székely discovered Radoszlav's treachery, he wasted no time in confronting him. György, whose skills with the sword were legendary, knocked Radoszlav off his horse and decapitated him in front of his unit. And while he was busy confronting the turncoat, events near Kolozsvár took a decisive turn.

જ

"There comes Lénárd," opens Taurinus's account of the Kolozsvár conflict,

> the pride of this Barlabási clan,
> good nobleman he is, vice voivode,
> leading the brave Szeklers into the fight,
> and the well-armed Saxons follow him.

At this point, Taurinus's humanist poem abandons the flowery prose and allusions to Greek mythology. He instead details the events, an account corroborated by many independent sources. His accuracy suggests that even if he was not eyewitness to what transpired, he must have at least received firsthand accounts from those who were.

Taurinus writes of the other Transylvanian vice voivode, János Thornallai, as well, whose fresh grief has turned to rage over the loss of his son, Miklós, hanged two months earlier at Nagylak on György's orders. Our poet speaks of János Bánffy, a rich landlord driven from his home by Lőrincz's men, now joining the noble throng. There was János Drágffy, whose presence on the battlefield is later substantiated by the Transylvanian bishop's surviving letter, in which the man of God orders his bailiff to round up as many serfs from the episcopate's land as can be found and march with them to Drágffy's aid. And there are the Szeklers of András Lázár, joined by the Saxons of Kronstadt and local Romanians. All stand united in their desire to end Brother Lőrincz's Transylvanian Crusade.

~

Meanwhile, at Temesvár, Voivode Szapolyai took advantage of the chaos caused by the infighting between György and his Serbian captain to launch his own assault. In short order, however, he understood that, despite the turmoil and mutiny among the serfs, the battle would not be so easily won as he had promised his men. The ferocity of the resistance surprised him, and for hours on end of bone-crunching battlefield horrors it seemed that the victory could go either way.

Szapolyai watched in increasing dismay as the lightly armed outlaws trampled over one another in their eagerness to fight his men. His infantry slowly lost ground, under pressure from the vicious onslaught. His cavalry, stuck fast in the deep, bloody mud of the battlefield, had been unable to take advantage of their height and speed to turn the tide. With that, the fortunes of the battlefield slowly but visibly tilted toward the numerous crusaders.

Amid the deafening clash of iron on iron, his decision to split the forces on two different fronts began to haunt the young voivode. If the Szeklers and the Saxons now in Kolozsvár had instead fought at his side, the overwhelming force would have guaranteed victory. With each passing minute their absence was more painfully felt.

~

Meanwhile, back at Kolozsvár, Brother Lőrincz's command had to have been crippled after the judge double-crossed him, detaining several of his captains. Otherwise, how can we explain Taurinus's verse, which hints at an ambush rather than the clash between the two battle-ready armies?

> *The peasants fought for their camp between the tents,*
> *"In arms! In arms!" they shouted in unison.*

As one would expect, Taurinus, Cardinal Bakócz's former secretary, captures the drama from the perspective of the nobility. We learn that Drágffy's horse stumbles after its leg is cut by a peasant's scythe, and we follow Bánffy and Thornallai as they rush to his side, carrying Drágffy off to safety after a hard-fought rescue. With that, they became the heroes of the Kolozsvár battle.

Such is typical in chronicles of that day—peasants are faceless, nobles are heroic, and their deaths are grisly. As the nobility won this battle, all of Taurinus's main protagonists managed to survive. Only one person, by now familiar to us, did not outlive the Kolozsvár battle: István Telegdi. We shall return later to him.

Under Brother Lőrincz's skillful command many of the defeated crusaders escaped, retreating in an orderly manner. Thus the lords' victory was not so much overwhelming as it was definitive, since the crusaders thereafter failed to rebuild the lost strength. Those captured bore the brunt of their foe's terrific rage, as recounted in Taurinus's verses:

> *The serfs left alive, they were all gathered*
> *and thrown in dark prison cells, where guarded till killed*
> *in selected manners: Some fell under the hatchet, others impaled;*
> *many turned into ashes on huge stakes,*
> *horrible how their blood sizzled when it dripped in the fire.*

Only ashes were left and bleached bones,
partially smoldered bodies tied to the stake,
a gruesome sight as they dried in the hot air.

❧

Back in Temesvár, what never came to be is just as important as what did. An accepted tactic of medieval warfare dictated that if a liberating force were to come to the aid of a sieged fortress, the fortress's defenders should leave the protection of their walls, pressuring the enemy from both sides. Inexplicably, the chronicle of Istvánffy is convinced that Báthory disregarded this custom and instead followed from the safety of his walls as the crusaders leveled the Transylvanian forces. On the other side of the battlefield the voivode watched helplessly as his famed hussars fell before the savage peasants. As a last resort he engaged his reserves, sending the palace guard, the mercenaries, and the restless Szeklers into the fray.

Szapolyai's contemporaries insist that "there was nothing remarkable" about the Szeklers' military supply, as "they lacked any special equipment, weaponry, or other military decoration." Even so, they flew to the heart of the battle like poisoned arrows. It was not superior weaponry but their "faith in their old and magnificent bravery that helped them fight till the end," the medieval chroniclers tell us. Sure enough, soon after they joined the battle the tide turned—decisively, lethally—in their favor. As Taurinus recounts, "the peasants wavered a bit and gave up their positions and at the end panicked and fled from the battlefield."

Only the outlaws surrounding György Dózsa Székely continued to fight, defending their captain with a frenzied resolve. But with the reatreat of the two crusader flanks, the captain and his protectors were exposed to the Transylvanian forces on all sides. Peter Petrovics, a young knight and distant cousin of the voivode, took advantage of the broken ranks and closed in on György. In the heat of the fight he managed to push the captain off his horse and captured him alive. Almost simultaneously, György's brother, Gergely, was also taken prisoner.

The loss of their leaders was the final blow to the peasants' resistance, and they scattered across the battlefield. Enraged as they surveyed the scores of fallen comrades before them, shaken from the victory that had been far too close for comfort, the knights mercilessly rode down the escaping peasants. And with that, the battle at Temesvár, just as at Kolozsvár, closed with a massacre of sickening proportions.

As we lament the Crusade's sudden dissolution, the brutal end of a peasant uprising in protest of the corrupt feudal system, let's note the tremendous irony: György Székely, a Szekler by birth, rose quickly from his Transylvanian obscurity. Without premeditation, he nearly succeeded in establishing the Szeklers' culture of freedom in Hungary. Just as he and his peasant forces were poised to prevail over Szapolyai and his Transylvanian army, the only force left to stop György accomplish his dream, he was crushed by a small unit of Szekler warriors, his kinsmen, thrown into the fray only in the battle's final, fevered minutes.

≈

And then, according to some historians, there wasn't any battle at all. Rather, when the voivode called for a peaceful surrender, many peasants immediately complied. In this version of events Radoszlav is still shown no mercy. In the midst of the chaos caused by the infighting, Peter Petrovics captures György, ending the battle before it even really begins.

Still other sources maintain that Petrovics seized György while the captain was on a reconnaissance mission, trying to size up the Transylvanian army and its positions. In this version of the events the peasants disperse without a fight after their leader is taken prisoner.

King Ulászló's letter of July 24, 1514, corroborates the last account; in it he writes to the German emperor that György Dózsa Székely was captured, and with that, "under Temesvár the peasants were calmed and dispersed without bloodshed." Many historians suspect that he might not have been telling the whole truth. Eager to downplay the ambitious Szapolyai's role in quelling the civil war that he himself and the cardinal had incited, the king may have fudged the importance of the Temesvár battle.

No matter which version we choose to believe, all accounts agree on certain key details: No one disputes that György Dózsa Székely and his reluctant brother, Gergely, survived, both taken prisoner. In Kolozsvár, Brother Lőrincz avoided capture and continued his fight for freedom elsewhere in the kingdom. And, this being Transylvania, as long as our characters are still drawing breath—and sometimes even after—our story cannot end just yet.

27

The Third Ear

I knew at least *one* person at just about every conference I had attended in the past. Some of the speakers, an organizer, or any number of the participants—there was always somebody familiar. But on November 9, 2006, I was out of luck as I scanned the casually elegant crowd of about a hundred assembled for cocktails at the mayor's official residence in Prague. So, armed with a glass of wine, I began practicing the ancient art of mingling, until I settled into conversation with a man in his early sixties.

Given the din of conversation all around us, I didn't catch his name. He was short and conventional-looking, a Mediterranean type, pairing a black leather jacket with a black Nike shirt. After a few minutes' idle chitchat, I offered the question always handy for oiling a conversation: What do you do?

He was an artist, of course, like just about everyone else in the Art Deco salon. So I quickly refined my question: What *kind* of art do you do?

For a split second, his face flashed puzzlement. As I later learned, he was the keynote speaker of the conference that had brought me to Prague in the first place. But my plane had landed three hours late, and I had missed his talk, making me perhaps the only person in the entire crowd who had no idea who he was. After that almost imperceptible hesitation, he gamely decided to invest in my education.

"This is what I do!" he said, and with his right hand he began pushing the left cuff of his jacket up his arm, slowly revealing his pallor. Then, about four inches from his elbow, the white skin's smooth flow was abruptly interrupted by . . . an ear.

Not a tattoo and not a prop glued to his arm, but an ordinary human ear covered with skin the same pale shade as his forearm.

I involuntarily glanced up and quickly confirmed he did, in fact, still have both of his ears on his head. And a third one, a perfectly normal-looking one, graciously resting on his left arm.

I got it now. His skin was his canvas, and the ear, the extra one, was his art.

That same night at a dinner I got a chance to confirm the authenticity of the ear—I even touched it, and it surely did feel real. I also learned his name: He was Stelarc.

Stelarc and nothing else.

This was the only name printed on his passport.

And the third ear was by no means his sole claim to fame. Rather, his suspension series had earned him initial stardom some decades earlier. He had typically been naked for these performances, hanging either from a series of cables in Tokyo or suspended facedown over East Eleventh Street in Manhattan.

It wasn't his nudity that stopped the passersby in their tracks. Or, at least, it wasn't *only* his nudity. It was the way he was suspended in midair. Several meat hooks pierced his skin from shoulders to kneecaps, keeping him aloft. He was hooked to art, his skin tugging and stretching as the pulleys lifted him into the air.

For his latest performance, he had contemplated the surgical removal of his ear and its transfer onto his forehead. His doctors had talked him out of it, assuring him that the transplant would probably die, costing him both his ear and his masterpiece. So they gave him an extra ear instead.

A soft prosthesis had been inserted under his skin, which he'd had to inject daily with a saline solution. After two months, the skin had stretched into a bubble, allowing the surgeons to implant an ear-shaped structure under it. He was then injected with stem cells, which grew into the ear's cartilage.

As our dinner chat drifted from his extra ear to the keynote address on human mobility I was to give later during the conference, we broached the topic of privacy. This is always a sensitive issue, and I braced myself for the extreme reaction I'd come to expect. Stelarc's outburst took me by surprise, in fact, but not for the reason I expected.

"I don't think we have too much surveillance," he said. "On the contrary, we have too little."

Startled, I pressed him to explain himself, and he gladly obliged. He thought that we need more and better detectors, not only around us but also within us. They should monitor everything, from our activities to every change in our bodies.

"I want to know when something in my cells goes wrong," he said. He wanted little robots inside his veins to detect anything that could make him sick. They would be his inner Vast Machine, arresting cells before they could become cancerous or blasting away clots in his bloodstream before they could give him a stroke.

Surveillance? As I soon learned, the extra ear was meant to be a tool for self-surveillance. It wasn't yet, and there was only one reason for that: His third ear was still deaf.

He hadn't intended for it to be that way. He'd had it implanted with a tiny microphone, but his skin had become infected, and the implant had to be removed. He was not ready to give up, however, and, as of the time of our conversation, was planning to try again. And still again, if need be. Once the mike was in place, it would be linked to the Internet so that anyone visiting his Web site could eavesdrop on him in real time. "Instead of an ear that hears, it will be an ear that transmits," he explained, making his self-surveillance public and instantaneous.

Stelarc's ear may not be functional yet, but his call for more surveillance is being rapidly fulfilled. And an unexpected side effect of this data-rich society in which we find ourselves has been the unprecedented detailing of our lives. Along the way a nascent science has begun quantifying everything we do, forcing us to rethink many things we take for granted, from free will to our privacy.

From my perspective, one of the most important findings of this new science is this: When we break our lives down to numbers, formulas, and algorithms, we are actually much more similar to one another than we might be ready to admit. Granted, you do what you think is best for you, and you do it when you can and when it is the most convenient for you. You may live in L.A. and I in Boston, you may be Asian and I Hungarian, you may run a restaurant while I do research, teach, and occasionally write books. These differences do matter; no one has ever really questioned that. But if we focus on our actions and their timing, we see patterns that are not unique to you and me. They pertain to billions. We are simultaneously bursty and quite regular. Apparently random yet deeply predictable. Sure, some events we face are quite haphazard. But the way we cruise through them is universal.

~

You may have been wondering off and on for some time: What is György Székely, this hero—or criminal—from the sixteenth century, doing in a book about human dynamics? Why mix science with history?

On the one hand it is obvious, perhaps: He was a burst that burned its fuel too fast. Barely three months passed between his rise to prominence and his capture, a brief moment from an historical perspective. But what months they were! He came from the bottom and virtually reached the top. Some believe that he could have ascended to the throne—others are convinced that in fact he did.* His story is vivid proof that when it comes to bursts we must not think only in terms of e-mail, the Web, and all things digital but must consider most aspects of life and history, from our diseases to battles fought in the sixteenth century.

But bursts aren't the only reason we have revisited György and his

---

* Many historians speculated that if, after avenging his fallen outposters at Nagylak, György Székely had turned his army against the defenseless Buda rather than the fortress Temesvár, he could have taken the king hostage and made history quite a bit more interesting. It appears that it was his deep respect for the crown that stopped him—he and his crusaders insisted to the end that they were loyal subjects of the king and that their only quarrel was with the landowning aristocracy. Nevertheless, there were many rumors after the fact that his crusaders had actually crowned György Dózsa Székely king.

bizarre Crusade. Telegdi was just as important for our journey. His prediction raises the possibility that our future, bloody though it may be, need not remain a mystery. If Telegdi could foresee the outcome of the Crusade in the sixteenth century, shouldn't we, half a millennium later, with science at our service, easily outdo him?

Even our famed philosopher naysayer Karl Popper acknowledged that long-term prophecies are possible for "systems [that] can be described as well-isolated, stationary, and recurrent." The planets and stars fall into this category; this is why we can foresee their trajectories. Yet "these systems are very rare in nature, and modern society is surely not one of them," Popper continued, and, with that, drove a stake once and for all through our attempts to predict our own futures.

So the real question is this: Which of the two prophets should we choose to believe? Is it Sir Karl, whose 1959 essay, built on the unforgiving logic of philosophy and science, insists we will never be able to peer into our future? Or do we listen to Telegdi and his prophecy validated by the unforgiving facts of history?

True, today we are nowhere near making the kind of long-term predictions that Telegdi offered his contemporaries. You may even argue that his was not even a prediction but, rather, something we all do when we are engaged in an argument—we build on our expertise and life experiences to call the outcome of future events. No one would confuse this with science—it is what it is, an informed opinion at best. If we are wrong, it is forgotten; if we are right, it does not make us a prophet.

There is an important difference, however, between predicting history, which is a collective mode of society, and predicting our daily activities. We may not be as regular as the planets, but many of our daily actions are quite repetitive in nature, making them foreseeable. So while prediction at the societal level remains cloudy at best, at the individual level it is growing increasingly feasible.

And it isn't merely our mobility that is regular enough to be predictable. Netflix and Amazon forecast our shopping preferences, banks calculate our financial trustworthiness, and insurance companies predict our chances of being in a car accident. A number of recent books like Ian Ayres's *Super Crunchers* and Stephen Baker's *Numerati* have documented how data mining exploits our deep-rooted predictability, changing every-thing from business to health care.* Indeed, everywhere we look, tech-nology stares back at us, transforming our desires and needs into dollar bills. Shopping. Travel. Entertainment. Love. Diseases. Relationships. Charity. Work. Everything we do is fundamentally affected by this per-vasive data mining. Stelarc's dream isn't so very far-fetched after all—we might soon carry tiny devices in our bloodstream as well.

No one expects these predictive tools to be perfect. How on earth could anyone ever have predicted that, despite my jet lag, I would choose to attend a cocktail party in Prague and that out of the scores and scores of guests the one I decide to chat with would have three ears? No, regard-less of how hard we try to keep to our well-trodden paths, we are bound to hit some bumpy anomalies in the road. Remember our friend Daniel, from back in Chapter 21, who usually shuttles back and forth between home and the office, day after day? Not even he was fully predictable if he even once decided to visit a pub with friends after work. His one, tiny, seemingly insignificant variation from routine increases his entropy, making him impossible to nail down with certainty.

But no physicist has ever successfully predicted the trajectory of 10^{23} molecules in a gas, either, and that hasn't stopped us from predicting the gas's pressure and temperature—arguably far more important than the trajectory of each individual molecule. The same is true for human dy-namics. Our deep-rooted unpredictability does not need to bubble up at the level of the society. If we carefully distinguish the random from the predictable, we might be able to foresee many features of the social fabric.

As we ponder this porous boundary between prediction and random-ness, we must realize that despite Popper's authority and impact, it is by no means obvious that he was right. Notwithstanding his claim, there is no solid proof that social systems are impossible to predict. Given that, the future holds two possibilities: It may usher in a new Heisenberg who

---

* The difference between human dynamics and data mining boils down to this: Data mining predicts our behaviors based on records of our patterns of activity; we don't even have to un-derstand the origins of the patterns exploited by the algorithm. Students of human dynamics, on the other hand, seek to develop models and theories to explain why, when, and where we do the things we do with some regularity.

will present a new uncertainty principle, telling us that Popper was right and that seeing into our future is not merely difficult but fundamentally impossible.

The alternate possibility is this: Driven mainly by commercial interests, predictive tools will continue to improve, particularly those that quantify individual behavior. And to further enhance their accuracy, these tools will move from focusing on individuals to focusing on the groups they belong to, as when you deviate from your regular pattern (head out to the pub instead of going straight home) it is often your friends who are to blame. The reach of these predictive tools will also extend from minutes to hours, which is a believable leap, given the short-term inertia in our actions. As the instruments struggle to expand from predicting hours to predicting days, they will at first be inaccurate, just as weather prediction was decades earlier. But their predictive power is bound to improve, until we may realize one day that our futures have ceased to be the mystery they once were.

꜔

Though it may appear that this story is nearing its end, it's really only about to begin. We are at a moment of great convergence, when data, science, and technology are all coming together to unravel the biggest mystery yet—our future, as individuals and as a society. Along the way György, Hasan, and the omnipresent bursts have helped us understand that the quest to unravel human dynamics is more than an intellectual exercise. It is the last pillar of science, and its impact may one day prove to be on a par with the physics of the early twentieth century or the unfolding revolution in genetics.

Scientists are poor prognosticators, often unable to see the consequences of their work. Luckily we have engineers and entrepreneurs to fill the gap between theorems and products. I am no exception to the rule—many companies have successfully monetarized my research on networks, none of which I foresaw. The same is true for human dynamics—I am too shortsighted to fully comprehend the full potential in forecasting your activity.

Lewis Richardson's first book was a remarkable failure. And yet today weather forecasting is routine, following the principles he outlined while driving ambulances on the battlefields of France. In his second book, *The Statistics of Deadly Quarrels,* he ventured farther, aiming to predict conflicts and wars in the hopes of helping us one day avoid them altogether. He failed once again. And so the real question is this: Have we matured enough to trust our predictive abilities?

Had the cardinal not been so fixated on currying favor in Rome, he might have listened to Telegdi, avoiding the ill-fated Crusade in its entirety. Instead, he pressed ahead single-mindedly, and in his wake thousands, serfs and nobility alike, were slain but no Ottoman lost his life. He also poisoned the trust between the classes, so that the peasants never again aided in the country's defense. Indeed, twelve years later at Mohács the nobility was forced to face the Ottomans alone, and lost not only the battle but also its king and the country with it. In the vacuum of power that followed, the Hapsburgs took Bohemia along with western Hungary, and Transylvania emerged as a semi-independent principality. What was left of the country crowned a new king, who then ruled as John I of Hungary. He had been known before as János Szapolyai.

Three years after the Kolozsvár battle Luther nailed his Ninety-Five Theses on the door of All Saints' Church in Wittenberg, and the Protestant Reformation swept through Europe. The dissent against the authority that the Crusade camps had nurtured paved the way for the Reformation's mass acceptance in Eastern Europe. The religious turmoil that followed tested Transylvania as well, prompting it to pass the first law ever proclaiming religious tolerance.

"In every place the preachers shall preach and explain the Gospel each according to his understanding of it, and if the congregation like it, well, if not, no one shall compel them for their souls would not be satisfied, but they shall be permitted to keep a preacher whose teaching they approve," said the Act of Religious Tolerance and Freedom of Conscience, issued in 1568 by the son of János Szapolyai, who followed his father as King John II of Hungary. This was two centuries before the 1786 Virginia Statute for Religious Freedom, and preceded by 131 years the 1689 English Act of Toleration. It also marked the birth of the Unitarian church, whose founder, Francis Dávid, a native of Kolozsvár and four at that time, must have watched the final battle between Father Lőrincz's peasants and Lénárd Barlabási's army with a combination of fear and wonder. So in the end the selfish actions of Cardinal Bakócz not only spelled the end of Hungary as an independent country, they also diminished, quite definitively, the church whose helm he had so coveted.

Are we fated to be forever at the whim of leaders whose private priorities drive their—and hence our—next moves, dragging us time and again into bloody quarrels? I heartily hope not. I hope that instead one day Lewis Richardson's dream will be fulfilled in that the pursuit of knowledge will guide our priorities. So in the end the predictive power of human dynamics is far more than credit scores and number crunching. It is a journey driven by the hope that one day it will make our planet a better home.

Which is a fine place to end, except that we have some unfinished business. I promised that we would reconsider Telegdi and his prophecy. I also want to check in with Hasan Elahi and where he stands currently with Homeland Security and their predictive algorithms. And of course there is György Dózsa Székely. What of his fate and that of his brother, Gergely, after their Temesvár capture?

First to Telegdi. György and Hasan will have to wait until our final chapter. But I warn you, you may wish to stop reading here, as some of what's about to unfold is deeply unsettling. And yet, knowing that what we are witnessing is entirely true, watching it unfold is fascinating as well.

❧

In the crumbling court of the Hungarian king, István Telegdi had thundered in opposition to the cardinal's Crusade: "I am sure that many peasants will be quick to assemble when summoned. But might they join the host merely to avoid the backbreaking effort to work the land or to avenge their many injustices suffered or to escape punishment and torture for crimes committed?"

But his words fell on deaf ears.

Looking back, his foresight seems extraordinary—many peasants, driven not by religious zeal but by economic hardship, joined the Crusades, just as Telegdi had foreseen. Many outlaws joined the host as well, since the camps offered them safe haven from prosecution for real or perceived crimes.

"And if, then, the nobility complain that the fields have been left to the weeds," he also said, which is precisely what the nobility did do, "this mob, shoddy crowd, now in arms, will you be able to keep them on mission?"

But just as startling is his conclusion: "The sword, given to them to destroy the enemy, will they not turn it against us . . . ?"

Again, in the light of the events we have witnessed, Telegdi's prophecy is astonishing.* Only now, five centuries later, have we started to make progress in predicting human activity. Yet, our ability to foresee wars and uprisings remains just as questionable today as it was for Lewis Richardson in the 1940s. Karl Popper proclaimed with thundering authority that "for strictly logical reasons, it is impossible for us to predict the future course of history." That is, foreseeing our future is not merely difficult or shady. It is downright impossible. Well, Sir Karl, how did Telegdi do it, then?

Humanist chroniclers learned to tell their tales by studying the ancient Roman historians, who had viewed their trade as literature, not history or science. That is, not only had they no qualms about inventing poetic words to put in the mouths of long-dead heroes, but they were also convinced that, given the need to reproduce the flow of history, doing so was their responsibility. Father Tubero, a Croatian friar who wrote his chronicle between 1522 and 1527 on the southern Adriatic island of Mljet, drafted an eloquent speech for György Dózsa Székely. Gianmichele Bruto in 1580 composed for him as many as three monologues.

---

* One thing is clear: Telegdi's oracular powers didn't save his life. Szerémi's history puts Telegdi among the Nagylak prisoners, arched to death when Bishop Csáky and other nobility met their end. We know, however, that despite the honorable intentions of our humanist priest, his words are not always trustworthy. Telegdi's alleged fate is one of those events that force us to question his accuracy.

On July 3, 1514, János Bánffy, the Transylvanian knight whose heroism we will witness a few weeks later at Kolozsvár, confronting the perils of the upcoming war, decides to pen down his will. Following contemporary custom several aristocrats witness his testament, János Telegdi among them. Dated more than a month after György Székely executed his prisoners, the will offers evidence that Nagylak did not see the end of our prophet. He did not outlive his signature by many weeks—a note found in Vienna indicates that within the next ten days he was captured in Kolozsvár and executed at the order of Brother Lőrincz.

In Istvánffy's widely read chronicle written around 1605 both György and Brother Lőrincz give extensive speeches, uttering words that went on to be quoted for centuries.

The truth is that nobody took notes in the court of Buda Castle when the cardinal outlined his plans for the Crusades and Telegdi spoke against them. We know of Telegdi's prophecy thanks to Szerémi and his influential chronicle of medieval Hungary. Aptly titled *Epistola de perditione regni Hungarorum*, it was written between 1545 and 1547, or about three decades after the events.* By then the outcome of the Crusades was no mystery to him or any of his contemporaries.

Telegdi's opposition to the cardinal's plan is probably historically accurate. Yet his prophecy owes more to the imagination of a humanist priest than any words he actually uttered. There is a lesson in this, one that continues to haunt historians and scientists alike: Nothing is easier than predicting the past.

---

* *Epistola de perditione regni Hungarorum* translates to "Epistle of the perdition of the Hungarian kingdom."

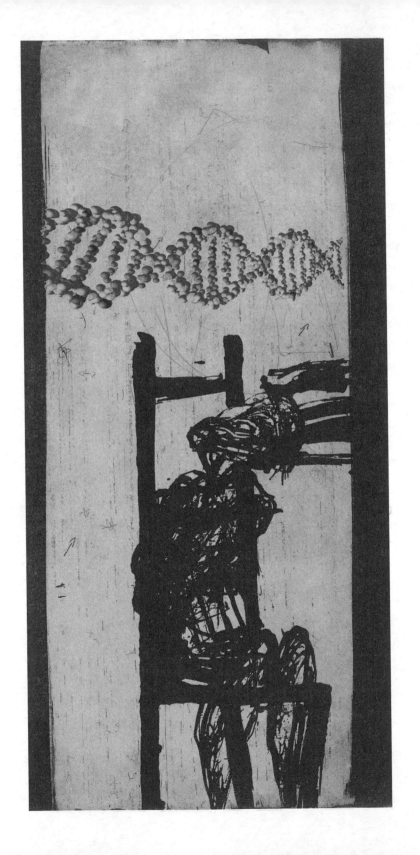

28

Flesh and Blood

 he swamps surrounding Temesvár had not fully dried yet, so it was hot and humid around the fortress. Nevertheless, the crowd was gathering, the boys first, followed soon after by the apprentices of the once-prosperous guilds—young men with little to do, as commerce was still in shambles. All that remained after the siege was free entertainment, and even the more respectable couldn't stay away—butchers, smiths, bakers, tailors, the town's rich and prosperous, were all on their way. Rich or poor, young or old, knight or civilian, they all had the same question on their mind: What will they do to him? How will he die?

A procession of brothers arrived, and the crowd parted to let them pass. They assumed the best spot on the field and watched the executioners pull pieces of iron from a white-hot fire. As the pieces started to fit together, they slowly recognized what was being prepared: The executioners had *forged a throne of iron**, as well as *an iron crown* accompanied

---

* The text in italics are quotes from original medieval sources describing the scene.

by *a scepter—the manifest insignia of any royal.* Would they execute or crown him?

A noble's blood was hard to spill—he might die in bed or perish in a fight, but it was rare to see one under the executioner's ax. Had the king perhaps pardoned him?

Of course, noble-born though he was, he had hanged a bishop and tortured knights to death. The crown could not forgive that. Savage though a war may be, it has its unwritten rules: You can torture peasants, outlaws, and criminals, but bishops and noblemen are untouchable. One guilty of violating these unspoken decrees must meet a particularly brutal end according to the logic of vengeance. Hanging is for criminals; beheading, swift and painless, is for disgraced noblemen; witches burn at the stake, and murderers are impaled. All of these deaths were too merciful for György Dózsa Székely. They must find something befitting his crimes.

Then the gates opened, and a small army left the fortress. First came the soldiers, followed by carts carrying about ten prisoners. They were not the mighty fighters the crowd had expected but broken souls being dragged to their deaths. Was it hunger that had subdued them? Rumor was that none of them had tasted a crumb since they'd been captured. Last among the prisoners were the two Szeklers, György and Gergely. Gergely was calm as he had always been, and György retained his dignity. At least that's how many would choose to remember him.

He never begged for mercy, but rumor has it he did voice one final wish: Spare the innocent, his younger brother, Gergely.

Would he get his way? The question created suspense among even the better informed. It could go both ways, they reasoned. Guilt by association demanded Gergely's death. Yet it was no secret that Gergely's voice of reason, while often ineffective, had saved the lives of many; perhaps this would earn him a reprieve. But when Gergely had appeared with the rest of the condemned, the debate was muted.

Executions were no somber events in those days but spectacles to be marveled at. With ten men served up that day, the viewing promised to be particularly spectacular. And among them was no ordinary criminal, either, but the man who could have ended their lives in a matter of days.

So they all would have had good reason to follow custom and hurl insults at the prisoners as they passed.

They were quiet, however, respectful in a way. These were not thieves and robbers but knights and soldiers, so an unusual mix of awe and fear enveloped them as they spotted the prisoners.

In silence the crowd watched the prisoners arrive, and the executioners took György to one side: *undressed him and set him on a tall chair,* the throne that had been pulled from the fire. Then they *placed the red-hot crown on his head and the scepter, glowing from the heat of the fire, in his hand.*

Some could not bear to watch the cruelty and turned their heads away. And some recognized the significance of the torturous death—knights who turn against their king have iron nailed to their heads. To most, less versed in royal symbolism, his chosen manner of death symbolized something the orchestrators had failed to consider: He was king in life and was to die like one—on his throne, scepter in hand, a crown on his head.

The red-hot regalia were not sufficient to snuff out his life. And that was just fine with his tormenters, who didn't wish to hurry his end. So *they then led the nine starving prisoners—by now pale as ghosts, on the threshold of death resembling more silhouettes than men—to György's "throne."*

This was the second act, the spectators all knew, and they trembled. What exactly was in store for them? No one dared ask, however. *And those present there, brothers, priests, and learned men, they all sang* in unison *Te deum laudamus.*

What an odd choice, perhaps. And yet the hymn fitted the occasion—a plea for mercy despite past sins, a gasping petition for union with the elect. But both the time and the place were mistaken.

Had the events unfolded as intended, György would still have expected the subdued soundtrack. But not here. Not now. It was to have been in Buda, in the Mátyás Church, while the banners of the vanquished, fallen during his victory at Constantinople, were hung. No one seemed to notice, however, this spatial and temporal dislocation, and the events continued their course: *In the midst of this singing, Gergely, his younger brother, who had been caught with him, was cut in three in front of him.*

The rest of the prisoners, would they share Gergely's fate? None expected otherwise.

But, no, they were not to be slain. At least, not yet.

The executioners removed several pliers from the fire and with vigor pushed the burning iron to György Székely's skin, pulling his live flesh between the pincers. Some were nauseated, and a few fell into a dead faint. Only the victim said nothing—resigned to a fate he endured with miraculous quiet.

Then *they ordered the prisoners to use their teeth to tear György's flesh from his body, sizzling and dripping where the hot iron had touched it, and to swallow it.*

While the sentence was horrific and surprising, the learned ones once again found the logic in the design. The country is a body, organic and alive, as the king had often emphasized. György must die the way he disfigured his homeland—by biting into its live body.

Some of *his servants resisted, three or four, and they were cut down unceremoniously. When the rest saw this, they jumped on him with open mouths and tore chunks of meat from his body.*

And again, those ignorant of the intended symbolism recognized something familiar and soothing: "Take, eat; this is my body," spoke the scene, while the Te Deum filled the heavy summer air. For many, this had ceased to be an execution and became instead a liturgy for a martyr.

He was still alive, however, and it was clear to those present that, despite his suffering, *György was not afraid. He neither cried nor moaned. He did vilify his torturers, however.* Dogs, he called them repeatedly, *Bastards whom he himself had raised.*

Over the centuries many have asked, Can anyone truly endure so much pain? Yes, one can, if he is not an ordinary man. György was a soldier of the Ends and a Szekler, strengthened by the morale of the hearty nomads and Huns, meting out death without mercy and compassion but willing to accept the cruelest end without a whimper.

Eventually, when the pain became too much, he breathed his last.

Then the serfs and peasants, once crusaders and now prisoners awaiting their fate, were all let go, unharmed. So were his disciples, those who had been willing to swallow his flesh.

And over the years the spot where György breathed his last became a site of pilgrimage. Indeed, witnesses swore that *Nagyboldogasszony* herself—the Virgin Mary—descended to lift him into Heaven.

~

Today the Renaissance castle on the Buda hill where György Székely was knighted and where Telegdi pronounced his prophecy is long gone. In its place is a Baroque palace, the home of the Hungarian National Gallery. If you follow its wide marble stairs up to the fifth floor, you will arrive under a large dome. There is a metal sculpture there, some ten feet high, iron and red copper welded into a startling sight. It's a man reduced to his skeleton—only a few of his muscles remain. At a first glance, you might think it is Christ, as you sense his suffering and pain. His mouth is frozen in a grimace, as if he is trying to gag out his agony.

The plaque affixed to the statue's base reveals little—*Tüzes Trón,* it says, followed by the name of its Transylvanian-born sculptor.* The lack of guidance is not unusual, forcing the visitor to decide whether his muscles were consumed by centuries of pain or, like some things made in Transylvania, he was perhaps reborn in this fleshless state. Contrasting his fragmentary whole, two accessories remain intact: the throne he sits on and his irregular crown.

Who is he? you, a visitor, may ask. But here any child would instantly know: György Dózsa Székely wears the crown, seated on his flaming throne. While terrifying, obscure he is certainly not.

Up until the nineteenth century most historians—nobility or priests with a romantic longing for bygone days—treated György as a despicable criminal. Those without a vested interest in preserving the class order begged to differ, however, a sentiment that over the centuries came to dominate. But it wasn't until the nineteenth century that historians started to give the man his due: György Dózsa Székely was not a daredevil, a monster, or an opportunist but a revolutionary with an outsized dream, fighting to change the prevailing order for the benefit of the impoverished masses. Suddenly a flood of novels, dramas, and poems took up his cause, and the history books were rewritten in his favor. Line by

---

* *Tüzes Trón* means *Flaming Throne.*

line, he grew larger than life, becoming the mythical creature he is today: the man who followed the destiny imposed on him by the accidental prophet János Telegdi.

❧

"Is today Tuesday?" The voice was calm—disturbingly so, like that of an executioner readying a lethal injection.

"Are we in Florida?"

"Is your name Hasan?"

He answered yes to each question. It's not like he really had a choice— it *was* Tuesday, he *was* in Florida, and his name *was* Hasan. Hasan Elahi, to be precise.

Almost swallowed by the old, poufy leather chair he sat in, Hasan stared at the striped wallpaper while countless wires monitored his blood pressure, respiration, and pulse. This was no doctor's office, and it wasn't a hospital either—his inquisitors were scanning for changes in his vital signs, signatures of deceit. This was the polygraph Hasan had agreed to take to end once and for all the ordeal that had started with his fateful stop in Detroit.

Finally, after the many dull questions came one that had some substance to it: "Do you belong to any groups that want to harm the United States?"

How am I to answer this? Hasan wondered. The response he was instructed to give, a simple yes or no, could not do the question justice. So far he had been completely truthful, but would his ambivalence now change the readout?

"Well, you know, I work at a university," he said finally. "I'm sure we've got some professors who are not exactly happy with the U.S. You might want to ask those guys."

Now, had he just admitted to being part of a group that wanted to harm the United States? Or was he just insisting on being funny again in a place where jokes had no place? The administrator of the polygraph did not miss a beat, but continued in his slow monotone: "Aside from what we have discussed, do you belong to any groups that want to harm the United States?"

Hasan's answer this time was a simple "No."

But this, too, passed, and Hasan was let go. On his way out he was greeted by his FBI agent, who told him that he had passed the polygraph and so was free to go home now. This was good news, of course, and, relieved that he finally had his life back, Hasan dared to ask, "Okay, can I get a letter saying that?" All he wanted was some proof of his innocence, as "at the end of the day, the chances of my being pulled back into this— it's small but it is real."

He never got it, of course.

"I knew that whatever was happening was outside the law," he told me without any bitterness in his voice. That is, for all practical purposes, Hasan was not a suspect, was never detained, interrogated, or harassed. Given that, no one could confirm or deny his guilt or innocence.

Hasan's extralegal state endured about seven years, until February 2009, when I noticed a new status posting on his Facebook page. It was brief but huge:

> Hasan Elahi now has a formal clearance from Homeland
> Security!!!

Wow, I thought, and fired off an e-mail right away, asking him, How did it happen? Why now? I was hoping to hear about an elaborate mea culpa, or at least a letter saying sorry, and perhaps a fat check from the government in recompense for the years of harassment he had endured. There was none of that. Instead, it had been plain serendipity.

Hasan had been invited to do a public art commission for the new terminal at the San Jose International Airport. In order to be paid, he'd had to be an airport employee, and each employee was required to undergo a "security-threat assessment" by the Department of Homeland Security. So, in the end, as Hasan put it, "in a roundabout way, getting a secure badge to enter sterile areas of the airport indicates that I'm not considered a security threat." Or at least not a great enough threat to be denied the badge.

In a passive-aggressive way, the government had admitted nothing and revealed nothing. They just let him be, as they had right after Detroit, countless times in Tampa, after the polygraph test in Florida, or at

JFK when he had learned by accident that his "suspicious movement" was a threat to national security. And Hasan can consider himself lucky that he was not born in the sixteenth century. Life was worth much less back then, and under similar circumstances he would probably have ended like György or his younger brother, Gergely. This being the twenty-first century, happily, he was given a fresh start that he embraced wholeheartedly.

Many of us, after being dragged in by the FBI, would probably lie low, avoid trouble, and hope that it would all blow over. Not Hasan. He ran with it, fashioning of his ordeal a mirror that he turned on our society. He channeled all of his fear and frustration into works of art that are witnesses to an era when suspicion was thick in the air. True, he gave up his privacy in exchange. Then again, we have all lost it anyway, haven't we?

The only difference between Hasan and me may not be that significant: His life is in museums and art galleries; mine is on password-protected hard drives. All I get is a false sense of security. Along the way, Hasan has helped us understand that the outliers are not only the celebrated ones—like the Einsteins, Picassos, and Gateses or the feared bin Ladens—but can be ordinary men with itchy feet and unusual daily habits. He's helped us understand that being different does not make you automatically a genius, or a terrorist.

So, despite his clearance, Hasan's project is not yet over. He now hopes to release his DNA sequence, offering Homeland Security the opportunity to isolate the terrorist gene among his three billion base pairs. They may not know as they begin their search that the gene they are most likely to find does not encode for terrorism. It controls our lack of predictability, and it is our best defense against the Vast Machine. I even have a name for it. Call it RNG. Or perhaps the DICE gene.

Notes

CHAPTER 1: The Best Bodyguard in the Business

While snippets of Hasan Elahi's story were told by numerous news outlets, the narrative reproduced in this book is based on a series of interviews that I conducted with him during 2009.

Gordon Bell's story is described in several books and articles. See, for example, *Total Recall: How the E-Memory Revolution Will Change Everything* by Gordon Bell and Jim Gemmell (New York: Dutton, 2009) or the article by Gordon Bell and Jim Gemmell, "A Digital Life," in the March 2007 issue of *Scientific American*. There are several third-party accounts of Bell's project, from *The New Yorker* ("Remember This? A Project to Record Everything We Do in Life" by Alec Wilkinson, May 28, 2007) to *Wired* magazine ("Microsoft Researcher Records His Life in Data," by Steven Leckart, August 24, 2009).

Deb Roy's work on recording his child's speech acquisition is described in Rony Kubat, Philip De Camp, Brandon Roy, and Deb Roy, "Total Recall: Visualization and Semi-Automatic Annotation of Very Large Audio-Visual Corpora." *Ninth International Conference on Multimodal Interfaces* (2007). See also "Self-Experimenters: Can 200,000 Hours of Baby Talk Untie a Robot's Tongue?" *Scientific American*, March 17, 2008.

CHAPTER 2: A Pope Is Elected in Rome

The election of Leo X is described in several books focusing on the history of the papacy. The list of the cardinals and the detailed results of the scrutiny are available in Ludwig Pastor, *The History of the Popes,* vol. VIII (St. Louis: B. Herder, 1908). It is also recounted in Cardinal Bakócz's biography, Fraknói Vilmos, *Erdődi Bakócz Tamás élete* (Budapest: Méhner Vilmos Kiadása, 1889).

On the overall process of papal election, see Frederic J. Baumgartner, *Behind Locked Doors: A History of the Papal Elections* (New York: Palgrave Macmillan, 2003); Michael Walsh, *The Conclave: A Sometimes Secret and Occasionally Bloody History of Papal Elections* (Norwich, Conn.: Canterbury Press, 2003); T. Adolphus Trollope, *The Papal Conclaves, As They Were and As They Are* (London: Chapman and Hall, 1876).

Note that the smell of Michelangelo's frescoes might not necessary have been a reminder of poor life—indeed, women used the same substance as painters for makeup, and the smoke of incense could have also suggested a rich and expensive rather than a poor environment.

For a detailed discussion on Vatican and Michelangelo's frescoes, see Marcia Hall (text) and Takashi Okamura (photos), *Michelangelo: The Frescoes of the Sistine Chapel* (New York: Harry N. Abrams, 2002), and Bart McDowell (text) and James Stanfield (photos), *Inside the Vatican* (Washington, D.C.: National Geographic Press, 1991).

CHAPTER 3:   The Mystery of Random Motion

There is a charming documentary about the Georgers and Lars Hufnagel, the creator of the WheresGeorge.com Web site, entitled *Wheresgeorge.com,* released in 2006 by Brian Galbreath Productions.

Dirk Brockmann's research on dollar-bill motion was published in D. Brockmann, L. Hufnagel, and T. Geisel, "The Scaling Laws of Human Travel," *Nature* 439 (2006): 462–465.

Dennis Derryberry continues to build and sell custom-made furniture in Vermont—see www.hastingshill.com.

Einstein's 1905 letter to his friend Conrad Hebrich is reproduced in *The Collected Papers of Albert Einstein,* vol. 5: *The Swiss Years: Correspondence, 1902–1914* by Albert Einstein (author), Anna Beck (translator) (Princeton, N.J.: Princeton University Press, 1995).

For an excellent discussion on the background and origins of the five papers Einstein wrote in 1905, see John Stachel, *Einstein's Miraculous Year: Five Papers That Changed the Face of Physics* (Princeton, N.J.: Princeton University Press, 2005).

On the life and accomplishments of Jean-Baptiste Perrin, see the obituary written by J. S. Townsend, in *Obituary Notices of Fellows of the Royal Society* 4, no. 12 (1943): 301.

CHAPTER 4: Duel in Belgrade

The Belgrade duel is a widely quoted moment of the 1514 events, being mentioned in many contemporary texts. When it comes to the historical background of the events recounted in this book, I have found particularly valuable the careful monograph of Gábor Barta and Nagy Antal Fekete, *Parasztháború 1514-ben* (Budapest: Gondolat, 1973), which takes a critical look at the events and the earlier sources, discussing the reliability of each account. A shorter, more popular version of the same material is Gábor Barta, *Keresztesek áldott népe* (Budapest: Móra Ferenc Könyvkiadó, 1977). An enjoyable account of the events is by István Nemeskürty, *Önfia vágta sebét, Krónika Dózsa György tetteiről* (Budapest: Magvető Kiadó, 1975). Finally, György Székely's detailed biography, Sándor Márki, *Dósa György* (Budapest: Magyar Történelmi Társulat, 1913), is also rather informative and rich in (often contradictory) details. As original sources, see the collection of documents pertaining to the Hungarian peasant wars, László Geréb, *A magyar parasztháborúk irodalma* (Budapest: Hungária Könyvkiadó, 1950). Another early take on the history of events is Géza Féja, *Dózsa György, Történelmi Tanulmány* (Budapest: Mefhosz Könykiadó, 1939).

The words of Ali of Epeiros are based on Ludovici Tuberonis, *Commentariorum de temporibus suis libri*, 1603, reproduced in Gábor Barta and Nagy Antal Fekete, *Parasztháború 1514-ben* (Budapest: Gondolat, 1973).

I have placed György's younger brother, Gergely, in the scene, despite the fact that some historians speculate that he joined György only later during the Crusades. This is plausible—no historical document places him at Belgrade.

CHAPTER 5: The Future Is Not Yet Searchable

"Fameiness" is the title word of a column by Meghan Daum, published February 17, 2007, in the *Los Angeles Times*.

On the modern-day reincarnation of Warhol's fifteen minutes of fame, see Josh Tyrangiel, "Andy Was Right," *Time*, December 25–January 1, 2007.

Collegium Budapest is a research institute in the Buda Castle that hosts the Institute for Advanced Studies. For more information visit their Web site, http://colbud.hu.

Our study on Web site visitation patterns is published in Z. Dezső, E. Almaas, A. Lukács, B. Rácz, I. Szakadát, A.-L. Barabási, "Dynamics of Information Access on the Web," *Physical Review E* 73, article no. 066132 (2006).

Note that recently Jon Kleinberg and collaborators have developed an automated method to track the news cycle and detect the burst of ideas and memes in blogs and Web sites. Their results are summarized in Jure Leskovec, Lars Backstrom, Jon Kleinberg, "Meme-Tracking and the Dynamics of the News Cycle," in *KDD [Knowledge Discovery and Data Mining] '09*, June 28–July 1, 2009, Paris, France, available on http://memetracker.org. The site offers an automated tool with which various memes can be tracked.

Isaac Newton's pursuit of alchemy is discussed in many sources. There is a special Web site devoted to it at Indiana University (http://webapp1.dlib.indiana.edu/newton/) and *Nova* has an hour-long documentary on the subject entitled *Newton's Dark Secrets,* which features the quote I used in the book together with a guide to the text by Bill Newman. For a biography of Newton, see James Gleick, *Isaac Newton* (London: Harper Perennial, 2003).

There are many sources detailing Rutherford's experiments that led to the discovery of transmutation. See, for example, A. S. Eve, *Rutherford* (New York: Macmillan, 1939); Norman Feather, *Lord Rutherford* (Priory Press Limited, 1940); Mark Oliphant, *Rutherford: Recollections of the Cambridge Days* (Amsterdam: Elsevier Publishing Company, 1972).

Should you be tempted to turn the remains of your loved ones into diamonds, you can do so by contacting www.lifegem.com.

CHAPTER 6: Bloody Prophecy

The speech of János Telegdi is from the chronicle of György Szerémi, *Magyarország romlásáról* [Georgius Sirimiensis: *Epistola de perditione regni Hungarorum,* written between 1545 and 1547], translated from the Latin by László Erdélyi and László Juhász (Budapest: Magyar Helikon, 1961).

King Ulászló was part of the Jagellion dynasty, which ruled Hungary, Poland, Bohemia, and several other central European kingdoms in the fifteenth and sixteenth centuries. For their rule and era, see Peter Kulcsár, *A Jagelló-kor* (Budapest: Gondolat, 1981).

CHAPTER 7: Prediction or Prophecy

For a biography of Lewis Fry Richardson, see Oliver M. Ashford, *Prophet—or Professor? The Life and Work of Lewis Fry Richardson* (Bristol and Boston: Adam Holger, 1985).

Richardson's 1922 book on weather prediction is now available as: *Lewis Fry Richardson, Weather Prediction by Numerical Process* (Cambridge: Cambridge University Press, 2007). For a modern take on weather prediction, together with a detailed discussion of the pitfalls and the promise of Richardson's method, see Peter Lynch, *The Emergence of Numerical Weather Prediction: Richardson's Dream* (Cambridge: Cambridge University Press, 2006).

For an annual list of failed predictions related to the business world, see the BTW section of *Business Week* each January. The prediction on the house prices by the Realtors Association comes from the January 8, 2007, and the January 12, 2009, issues of the magazine, pp. 14 and 15–16 respectively.

About the accuracy and pitfalls of predictions, see Philip E. Tetlock, *Expert Political Judgment: How Good Is It? How Can We Know?* (Princeton, N.J.: Princeton University Press, 2005).

Popper's argument against historicism comes from Karl R. Popper, "Prediction and Prophecy in Social Sciences," published in *Theories of History,* edited by Patrick Gardiner (New York: The Free Press, 1959), pp. 267–285. For a detailed critique of Popper's arguments, see Peter Urbach, "Is Any of Popper's Arguments Against Historicism Valid?" *British Journal for the Philosophy of Science* 29 (1978): 117–130.

CHAPTER 8: A Crusade at Last

While everybody calls it the Mátyás Church, after King Matthias, the Buda church's official name is Castle Church of Mary. For images of the church, see www.matyas-templom.hu. See also Balázs Mátéffy and Görgy Gadányi, *Living Stones: The Unknown Matthias Church* (Budapest: Viva Media Holding, 2003).

For a more detailed take on the historical Buda Castle, where many of the events recounted in this book took place, see László Gerő, *A Budai Vár* (Budapest: Panoráma, 1971); Zoltán Bagyinszki, *Magyar Kastélyok—Hungarian Castles and Mansions* (Debrecen: Tóth Könyvkereskedés és Kiadó, 2008); Zoltán Bagyinszki and Pál Tóth, *Magyar Várak—Hungarian Castles and Fortresses* (Debrecen. Tóth Könyvkereskedés és Kiadó, 2008).

Note, however, that it is not clear where the Mass described in this chapter took place. During medieval times the Mátyás Church was used mainly by the German community in the Buda Castle. Historians, from Barta and Fekete to Nemeskürty, follow the chronicle of Istvánffy (*Miklós Istvánffy magyarok dolgairól írt históriája* [Budapest: Balassi Kiadó, 2003]) and place the cardinal's Mass in the modest Zsigmond Chapel in the Buda Castle that once stood on the northern side of St. György Square. The Turks demolished the chapel, and its last remains were destroyed during the 1686 siege of Buda. The chronicle of Szerémi (György Szerémi, *Magyarország romlásáról* [Budapest: Magyar Helikon, 1961]), puts the event in the St. György Chapel.

The elements of the Latin Mass used in the chapter come from *The Roman Missal in Latin and English* (Collegeville, Minn.: The Liturgical Press, 1968). I have also consulted several other sources. See, for example, Dom Jean de Puniet, *The Mass: Its Origin and History* (New York: Longmans, Green and Co., 1939); A. Croegaert, *The Mass: A Liturgical Commentary* (Westminster, Md.: The Newman Press, 1958).

While I repeatedly call the 1514 historical events "medieval," we are in the midst of the Renaissance. Some historians would call the period postmedieval or early modern. Yet the socioeconomic conditions of Hungary and Transylvania were in many respects consistent with those of medieval times, despite the Renaissance architecture of King Mátyás, who brought Italian architects to Buda.

CHAPTER 9: Violence, Random and Otherwise

The story of Tim Durham has been told on numerous occasions. The most detailed account can be found in Barry Sheck, Peter Neufeld, and Jim Dwyer, *Actual Innocence: When Justice Goes Wrong and How to Make It Right* (London: Penguin Books, 2001). There is also a *Primetime Live* documentary by ABC News entitled "The Wrong Man," which aired on January 29, 1997. Tim Durham also appears in *The Innocents* (New York: Umbrage Editions, 2003) by Taryn Simon, featuring photos of exonerated individuals.

In reconstructing Tim's story I have also used newspaper accounts from the local newspaper *Tulsa World,* which followed the case closely, from the rape of Molly to the exoneration of Timothy Durham.

For a discussion on Poisson's life, see "A Portrait of Poisson" by B. Geller and Y. Bruk, *Quantum* (March–April 1991): 21–25.

For an English translation of Poisson's work relevant to the Poisson distribution, see S. M. Stigler, "Poisson on the Poisson Distribution," *Statistics and Probability Letters* 1 (1982): 33–35. To put Poisson's work in the context of the development of statistics and other probability-based works on the juridical system, see S. M. Stigler, *The History of Statistics* (Cambridge, Mass.: Harvard University Press, 1986).

For a more technical take on the mathematics and applications of Poisson distributions, see Frank A. Haight, *Handbook of the Poisson Distribution* (New York: John Wiley and Sons, 1967) and J. F. C. Kingman, *Poisson Processes* (Oxford: Clarendon Press, 1993).

An excellent mathematical exposition of Poisson's calculation is available in A. E. Gelfand and H. Solomon, "A Study of Poisson's Models of Jury Verdicts in Criminal and Civil Trials," *Journal of the American Statistical Association* 68, no. 342 (June 1973): 271–278. For the application of Poisson's ideas to the American jury system, see A. E. Gelfand and H. Solomon, "Modeling Jury Verdicts in the American Legal System," *Journal of the American Statistical Association* 69, no. 345 (March 1974): 32–37.

For the discussion on the merits of the smaller jury sizes, see Dennis J. Devine, Laura D. Clayton, Benjamin B. Dunford, Rasmy Seying, and Jennifer Pryce, "Jury Decision Making: 45 Years of Empirical Research on Deliberating Groups," *Psychology, Public Policy, and Law* 7, no. 3 (September 2001): 622–727. See also "12-Member Juries and Unanimous Verdicts: A Debate," *Judicature* 88, no. 6 (May–June 2005): 300–305.

CHAPTER 10:   An Unforeseen Massacre

Note that the evidence is somewhat contradictory with regard to when exactly the crusaders left Buda and when the Apátfalva battle took place. I have followed the timeline reconstructed by Gábor Barta and Nagy Antal Fekete in *Paraszthábarú 1514-ben* (Budapest: Gondolat, 1973). Combining the various pieces of evidence, they apply realistic travel times for a medieval army in order to arrive at a rational sequence of events.

CHAPTER 11:   Deadly Quarrels and Power Laws

While Richardson did publish his war-related findings in several scholarly articles, his best exposition on the subject is his book, Lewis F. Richardson, *Statistics of Deadly Quarrels* (Pittsburgh: The Boxwood Press, 1960), which was published only after his death in 1953. A more recent exposition on Richardson's results and contributions to war theory is in David Wilkinson, *Deadly Quarrels: Lewis F. Richardson and the Statistical Study of War* (Berkeley: University of California Press, 1980). For a shorter but quite delightful exposition, see Brian Hayes, "Statistics of Deadly Quarrels," *American Scientist* 90 (2002): 1015.

In the months leading up to the 2003 Iraq invasion, the proponents of the war repeatedly argued that, as there had never been a conflict between two democratic states, turning Iraq into a democracy would make this source of Middle Eastern instability into a beacon of peace. Once all countries elect a democratic government, the argument goes, wars will cease, and with that Richardson's Law should become moot sometime in the twenty-first century. It is ironic, therefore, that the Iraq war itself is today a case study for Richardson's Law. An analysis of the daily body count in Iraq indicates that during most days, there have been only a few casualties—yet these somewhat peaceful days must be measured against a few other days that have seen hundreds and occasionally thousands killed, rare events that are nevertheless the natural ingredients of the power law that all conflicts follow. See, for example, Aaron Clauset, Maxwell Young, Kristian Skrede Gleditsch, "On the Frequency of Severe Terrorist Events," *Journal of Conflict Resolution* 51, no. 1 (2007): 58–88.

The fact that an imminent worldwide peace is only an illusion was clearly demonstrated by a study that catalogued all conflicts resulting in thirty-two or more deaths between 1400 and 1999, including everything from the casualties of the 1514 Crusades to the 1995 Tokyo nerve gas attack. The number of violent conflicts showed no decreasing or increasing tendency during the six-hundred-year period, supporting Richardson's conclusion that wars are random in time, without a significant historical trend toward less or more violence. See, for example, Peter Brecke, "Violent Conflicts 1400 A.D. to the Present in Different Regions of the World," a paper presented at the 1999 Meeting of the Peace Science

Society on October 8–10, 1999, in Ann Arbor, Michigan; J. Alvarez-Ramirez, C. Ibarra-Valdez, E. Rodriguez, and R. Urrea, "Fractality and Time Correlations in Contemporary War," *Chaos, Solitons and Fractals* 34 (2007): 1039. For a potential mechanism behind the power laws, see N. Johnson, M. Spagat, J. Restrepo, J. Bohoquez, N. Suarez, E. Restrepo, and R. Zarama, "From Old Wars to New Wars and Global Terrorism" (preprint).

Note that Richardson's data was likely biased toward Christian conflicts, potentially explaining the abundance of such wars in his database. Indeed, it continues to be difficult to find reliable data on conflicts that have taken place outside Europe, thanks to the scarcity of sources and the lack of attention of Western historians to other regions of the world.

In the field of economics, quite a bit of work has been directed, during the past decade, at predicting the likelihood of conflict. These studies follow in the footsteps of Richardson but rely on much better data and tools. One prominent representative of the trend is Paul Collier, who uses probabilistic models to argue that economic causes correlate more with civil war onset than political causes.

Jean-Pierre Eckmann's paper on the Web structure, inspired by his search for revisionists, is: Jean-Pierre Eckmann, Elisha Moses, "Curvature of Co-Links Uncovers Hidden Thematic Layers in the World Wide Web," *Proceedings of the National Academy of Sciences (USA)* 99 (2002): 5825. His paper in e-mail communication is: Jean-Pierre Eckmann, Elisha Moses, Danilo Sergi, "Entropy of Dialogues Creates Coherent Structures in E-Mail Traffic," *Proceedings of the National Academy of Sciences (USA)* 101 (2004): 14333.

My paper reporting on the bursty nature of the e-mail pattern was published in A.-L. Barabási, "The Origin of Bursts and Heavy Tails in Human Dynamics," *Nature* 435, no. 207 (2005). The power law nature of the interevent times was independently observed by A. Johansen, *Physica A* 338, no. 1–2 (2004): 286–291.

On the power law interevent times in printing patterns, see Uli Harder and Maya Paczuski, "Correlated Dynamics in Human Printing Behavior," *Physica A* 361 (2006): 329–336.

The results on the timing of library visits were presented in A. Vazquez, J. G. Oliveira, Z. Dezső, K. I. Goh, I. Kondor, A.-L. Barabási, "Modeling Bursts and Heavy Tails in Human Dynamics," *Physical Review E* 73, article no. 036127 (2006).

The power law nature of the mobile-phone calling patterns is discussed in J. Candia, M. C. Gonzalez, P. Wang, T. Schoenharl, G. Madey, A.-L. Barabási, "Uncovering Individual and Collective Human Dynamics from Mobile Phone Records," *Journal of Physics A: Mathematical and Theoretical* 42, 2–11 (2008).

The power law nature of Web-browsing patterns and its application to the visitation patterns of a particular news item were discussed in Z. Dezső, E. Almaas, A. Lukács, B. Rácz, I. Szakadát, A.-L. Barabási, "Dynamics of Information Access on the Web," *Physical Review E* 73, article no. 066132 (2006).

CHAPTER 13:   The Origin of Bursts

The story on Ivy Lee and Charles Schwab comes from Mary Kay, *You Can Have It All: Lifetime Wisdom from America's Foremost Woman Entrepreneur* (Rocklin, Calif.: Prima Lifestyles, 1996).

Much has been written about Charles Schwab and his time. For Schwab's biography, see Robert Hessen, *Steel Titan: The Life of Charles M. Schwab* (Pittsburgh: University of Pittsburgh Press, 1975). I have also found a rather informative series of articles published in the December 14, 2003, issue of the Allentown (Pa.) *Morning Call* entitled "Forging America: The History of Bethlehem Steel." I could not identify its author, but the piece is available on the Web site of *The Morning Call*.

Quite a number of books on time management discuss the importance of priority lists. Those I consulted include Julie Morgenstern, *Never Check E-Mail in the Morning: And Other Unexpected Strategies for Making Your Work Life Work* (New York: Simon and Schuster, 2004), Alec Mackenzie, *The Time Trap* (New York: American Management Association, 1997), H. Eugene Griessman, *Time Tactics of Very Successful People* (New York: McGraw-Hill, 1994), Marshall J. Cook, *Time Management* (Avon: Adams, 1998), Julie Morgenstern, *Time Management from the Inside Out* (New York: Henry Holt and Company, 2000).

Csíkszereda, the town to which my parents moved when I started high school, and which subsequently became my hometown, is also known as the hometown of the Whiskey Robber, whose deeds are described in the highly entertaining book by Julian Rubinstein, *Ballad of the Whiskey Robber: A True Story of Bank Heists, Ice Hockey, Transylvanian Pelt, Smuggling, Moonlighting Detectives, and Broken Hearts* (New York: Little Brown and Company, 2004). For a detailed pictorial description of the city and its history, see Mária Vofkori, *Csíkszereda és Csíksomlyó képes története* (Békéscsaba, Hungary: Typografika, 2007).

The Indian conference I was traveling to was the International Conference on Statistical Physics, held between July 4 and 9, 2004, in Bangalore.

The priority model that explained the origins of burstiness in human activity patterns was published in A.-L. Barabási, "The Origin of Bursts and Heavy Tails in Human Dynamics," *Nature* 435, no. 207 (2005). For an exact solution of the priority model, see A. Vázquez, "Exact Results for the Barabási Model of Human Dynamics," *Physical Review Letters* 95, article no. 248701 (2005): 1–4. For a more detailed exposition on the problem, see, e.g., A. Vázquez, J. G. Oliveira, Z. Dezső, K. I. Goh, I. Kondor, A.-L. Barabási, "Modeling Bursts and Heavy Tails in Human Dynamics," *Physical Review E* 73, article no. 036127 (2006).

Note that recently several authors have proposed that burstiness can be explained within the context of Poisson's theory, as a variable-rate Poisson process. To the best of my knowledge, this explanation was first suggested by César A. Hidalgo R., "Conditions for the Emergence of Scaling in the Inter-event Time of Uncorrelated and Seasonal Systems," *Physica A* 369 (2006): 877–883. It was later expanded by R. Dean Malmgren, Daniel B. Stouffer, Adilson E. Motter, and Luis

A. N. Amaral, "A Poissonian Explanation for Heavy Tails in E-Mail Communication," *Proceedings of the National Academy of Sciences (USA)* 105 (2008): 18153–18158, and D. Malmgren, J. Hofman, L. Amaral, and D. Watts, "Characterizing Individual Communication Patterns," Paris, *KDD [Knowledge Discovery and Data Mining]* 2009.

Many aspects of the priority model have been studied subsequently in the literature. See, e.g., D. Helbing, M. Treiber, and A. Kesting, "Understanding Interarrival and Interdeparture Time Statistics from Interactions in Queuing Systems," *Physical A* 363, no. 1 (2006): 62; A. Vázquez, "Impact of Memory on Human Dynamics," *Physical A* 373 (2007): 747–752; P. Blanchard and M. O. Hongler, "Modeling Human Activity in the Spirit of Barabási's Queueing Systems," *Physical Review E* 75, no. 2, article no. 026102 (2007); A. Gabrielli and G. Caldarelli, "Invasion Percolation and Critical Transient in the Barabási Model of Human Dynamics," *Physical Review Letters* 98, no. 20, article no. 208701 (2007); C. Bedogne and C. J. Rodgers, "A Continuous Model of Human Dynamics," *Physica A* 385, no. 1 (2007): 356–362; G. Grinstein and R. Linsker, "Power-Law and Exponential Tails in a Stochastic Priority-Based Model Queue," *Physical Review E* 77, no. 1, article no. 012101 (2008); B. Gonçalves and J. J. Ramasco, "Human Dynamics Revealed through Web Analytics," *Physical Review E* 78, no. 2, article no. 026123 (2008); A. Grabowski, "Opinion Formation in a Social Network: The Role of Human Activity," *Physica A* 388, no. 6 (2009): 961–966; N. Masuda, J. S. Kim, and B. Kahng, "Priority Queues with Bursty Arrivals of Incoming Tasks," *Physical Review E* 79, no. 3, article no. 036106 (2009); B. Min, K. I. Goh, I. M. Kim, "Waiting Time Dynamics of Priority-Queue Networks," *Physical Review E* 79, no. 5, article no. 056110 (2009); D. Rybski, S. V. Buldyrev, S. Havlin, et al., "Scaling Laws of Human Interaction Activity, *Proceedings of the National Academy of Sciences (USA)* 106, no. 31 (2009): 12640–12645; J. L. Iribarren and E. Moro, "Impact of Human Activity Patterns on the Dynamics of Information Diffusion," *Physical Review Letters* 103, no. 3, article no. 038702 (2009).

Note, however, that the priority-based queuing model should not be applied to a single activity, e.g., *only* e-mail. Indeed, if we spend our whole day doing nothing else but writing e-mails, we will have a somewhat regular output, free of bursts. None of us writes e-mails twenty-four hours a day, however. There is news to check on, there are papers and books to read, meetings we need to attend, the dog that needs to be taken for a walk, the dinner that needs to be served, the poem that needs to be penned down, and on top of all this we do sleep, as well, occasionally. These tasks inevitably compete for our most valuable resource, time, forcing us to respond to our e-mails only when sitting down to the computer becomes a high enough priority. Even then our mind unconsciously makes a priority assessment, prompting us to respond immediately only to the important, timely e-mails, saving the less important ones for later. Thus the response time to an e-mail is decided by the particular message's priority, not compared to the other e-mails, but compared to all other tasks on our priority list. This is true for any activity, from phone calls to library visits: They are bursty because they compete with other tasks, generating the off-on-off pattern we customarily see in human activity.

For a recent review on human dynamics from the perspective of statistical physics, see C. Castellano, S. Fortunato, and V. Loreto, "Statistical Physics of Social Dynamics," *Reviews of Modern Physics* 81, no. 2 (2009): 591–646.

If you followed our argument carefully, you may have noticed that in the previous chapters we focused on the interevent time, representing the time between two e-mails sent by the same person or between two clicks on a Web site. In contrast, the priority model predicts the time each task *waits on our priority list*. But does our procrastination follow the same power law as the interevent times? It does, according to Jean-Pierre Eckmann. Indeed, he measured the time it took a typical user to respond to e-mails. Eckmann found that the vast majority of e-mails were dealt with right away, within five minutes to an hour of receipt. A few messages, however, had to wait days and weeks for a reply. The distribution of the response time followed the same power law as the interevent times, suggesting that procrastination drives the burstiness of our activity pattern. See Jean-Pierre Eckmann, Elisha Moses, and Danilo Sergi, "Entropy of Dialogues Creates Coherent Structures in E-mail Traffic," *Proceedings of the National Academy of Science (USA)* 101 (2004): 14333.

CHAPTER 14:   Accidents Don't Happen to Crucifixes

The scene with the crucifixes is mentioned in several sources, but exactly when or where the incident happened is not clear. Sándor Márki, in his Dózsa biography, places it at the Rákos field, next to Buda, citing as his source a document number 14.527 found in an archive in Vienna, Austria; but apparently other sources also mention the event, one dating it to May 25 (that is, two days *before* the date we used). In addition, granted the event did happen, the site could not have been near Buda, as according to later research summarized in the Barta-Fekete book, the cardinal's letter reached György Székely only after he and his army had left that city.

Given that the cardinal sent at least two letters to the crusaders, one ordering them to stop the recruitment and the other to end the Crusade, it is not clear which was accompanied by the crucifix event.

Note that even Márki's 1913 biography treats the story with suspicion. He suggests that the fall of the crucifix probably involved some human action from the Franciscan brothers who disagreed with the cardinal and favored the continuation of the campaign.

I used the historical novel of Eötvös József, *Magyarország 1514-ben* (Budapest: Magyar Helikon, 1972), published originally in 1847, as the source of György

Székely's speech, thus it is fictitious. Eötvös also places the scene on the Rákos field next to Buda.

CHAPTER 15:   The Man Who Taught Himself to Swim by Reading

The quote from Kaluza's son on swimming came from www-history.mcs.st-andrews.ac.uk/Biographies/Kaluza.html maintained by the School of Mathematics and Statistics, University of St. Andrews, Scotland.

For a detailed discussion on Kaluza's theory and the scientific background of his correspondence with Einstein, see Daniela Wunsch, "Einstein, Kaluza, and the Fifth Dimension," in *The Universe of General Relativity,* edited by A. J. Kox and Jean Eisenstaedt (Boston: Birkhäuser, 2005).

The letters exchanged between Kaluza and Einstein are available in *The Collected Papers of Albert Einstein,* vol. 9, *The Berlin Years: Correspondence, January 1919–April 1920,* edited by Diana Kormos Buchwald, translated by Ann Hentschel, with Klaus Hentschel as consultant (Princeton, N.J.: Princeton University Press, 2004). I have quoted from letters #26, 30, 35, 40, and 48, as numbered in the volume.

Our paper on Darwin's and Einstein's correspondence patterns was published in J. G. Oliveira and A.-L. Barabási, "Darwin and Einstein Correspondence Patterns," *Nature* 437 (2005): 1251.

Darwin decided to seclude himself from his time's academic establishment, so his correspondence played a key role in the shaping of his ideas. For an excellent biography of Darwin, see Janet Browne, *Charles Darwin: Voyaging* (Princeton, N.J.: Princeton University Press, 1995) and *Charles Darwin: The Power of Place* (Princeton, N.J.: Princeton University Press, 2002). See also David Quammen, *The Reluctant Mr. Darwin* (New York: Atlas Books, 2006).

Darwin's full correspondence has been published in a sixteen-volume series, extending from *The Correspondence of Charles Darwin,* vol. 1, *1821–1836* (Cambridge: Cambridge University Press, 1985) to *The Correspondence of Charles Darwin,* vol. 16, *1868* (Cambridge: Cambridge University Press, 2008). For more information on his correspondence, including access to many of his letters, see www.darwinproject.ac.uk.

Our model for the correspondence pattern is based on an old model from queuing theory, A. Cobham, J. Oper, "Priority Assignment in Waiting Line Problems," *Bulletin of the Operations Research Society of America* 2, no. 70 (1954).

For another example of power laws in the correspondence of famous scientists, see N. N. Li, N. Zhang, and T. Zhou, "Empirical Analysis on Temporal Statistics of Human Correspondence Patterns," *Physica A* 387, no. 25 (2008): 6391–6394, which explores the correspondence pattern of a Chinese scientist, Hsue-Shen Tsien. See also R. Dean Malmgren, Daniel B. Stouffer, S. L. Andriana,

O. Campanharo, Luís A. Nunes Amaral, "On Universality in Human Correspondence Activity," *Science* 325 (2009): 1696–1700.

In 1963 Sheldon Glashow, Abdus Salam, and Steven Weinberg proposed that the weak nuclear interactions, a force not yet known to Einstein, and electricity and magnetism could arise from a partially unified electroweak theory, landing them the 1979 Nobel Physics Prize. Yet, the grand unification that Einstein envisioned remains an elusive goal.

CHAPTER 16: An Investigation

Other than his scattered letters, there is little coherent information on Lénárd Barlabási. His biography is under preparation by Sándor Barabássy, tentatively entitled *Egy magyar reneszánszkori főúri mecénás család a 15.-16 századi Erdélyben, avagy Barlabássy Lénárd erdélyi alvajda, székely alispán kora és tevékenysége.* An early version of the family history by Sándor Barabássy, written in the 1970s, and still unpublished, has helped me in the writing of this book.

On January 11, 1870, some five weeks after the November meeting in Kolozsvár of the Hungarian Historical Society, the committee named Károly Szabó as the editor of the book series. The first volume of *Székely oklevéltár*, the product of Szabó's work, appeared in 1872. Károly Szabó published three volumes before his death in 1890. The project was restarted in 1983, with *Székely oklevéltár, Új Sorozat* (Bucharest: Kriteron Könyvkiadó, 1983). It was during this enormous project that Szabó found the document that shed light on György Székely's life before his emergence in Belgrade.

Károly Szabó's account of his find in Nagyszeben was published in Károly Szabó, "Dózsa György életére," *Századok* 20. évfolyam, 1-es szám (1876): 18–21.

The full transcript of the letter is reprinted (in Latin) in the third volume of the *Székely oklevéltár*, published in 1890. The full reference is *Székely oklevéltár, vol. III*, edited by Károly Szabó (Kolozsvárott, Nyomtatott Fejér Vilmosnál az Ev. Ref. Kollégium Könyv-és Kőnyomdájában, 1890), pp. 163–165 (document # 524). Its content was first revealed to me by József Darvas Kozma of Csíkszereda. A detailed English translation was provided by Daniel Perett. Note that I edited Daniel's translation to make it easier for the contemporary reader.

For those fascinated by Nagyszeben, I recommend the trilingual (Romanian, German, English) *Sibiu Hermannstadt* (Sibiu: Humanitas, 2007), published when Nagyszeben became the European Cultural Capital in 2007. See also Emil Sigerus, *Cronica orașului Sibiu 1100–1929* (Sibiu: Honterus, 2006).

On the lives and literature of the Saxons, see *Telepes népség—Erdélyi szász olvasókönyv*, edited by Zoltán Hajdú-Farkas (Csíkszereda, Hungary: Pallas-Akadémia, 2006).

CHAPTER 17: Trailing the Albatross

The quoted verses come from Taurinus's poem on the 1514 events (István Taurinus, *Paraszti háború*, translated by László Geréb, original title: *Stauromachia id est Cruciatorum Servile Bellum*, Vienna: 1519).

For a short biography of Paul Lévy, see Benoît B. Mandelbrot, *The Fractal Geometry of Nature* (New York: W. H. Freeman, 1983). See also J. L. Doob, "Obituary: Paul Lévy," *Journal of Applied Probability* 9, no. 4 (December 1972): 871–872.

A wonderful description of the ecology of random walks is found in "Beyond Brownian Motion," by Joseph Klafter, Mike F. Shlesinger, and Gert Zumofen, *Physics Today* 49 (1996): 33–39.

For a description of a life in the sciences under the Soviet government, and of the life of dissidents, see Mark Ya. Azbel, *Refusenik* (Boston: Houghton Mifflin, 1981).

The Boston University group's work on the Lévy flights in albatross flight pattern is published in G. M. Viswanathan, V. Afanasyev, S. V. Buldyrev, E. J. Murphy, P. A. Prince, and H. E. Stanley, "Lévy Flight Search Patterns of Wandering Albatrosses," *Nature* 381 (1996): 413–415.

One of the scientists who chose to test the Lévy paradigm was José Luís Mateos, a physicist collaborating with researchers studying the spider monkeys of the Yucatán Peninsula. One day he asked his colleagues if they knew the trajectory of the individual monkeys. They did, it turned out, not because they'd attached detectors to them, as Vsevolod had done with the albatrosses. Rather, each monkey was assigned a graduate student, who followed it everywhere, recording everything from its eating habits to its interactions with other monkeys. While the monkeys did not, the graduate students did carry GPS devices, allowing Mateos to reconstruct their path across the forest, offering evidence that Lévy flights captured the monkeys' trajectories. For a description of their results, see Gabriel Ramos-Fernández, José L. Mateos, Octavio Miramontes, Germinal Cocho, Hernán Larralde, Barbara Ayala-Orozco, "Lévy Walk Patterns in the Foraging Movements of Spider Monkeys *(Ateles geoffroyi),*" *Behavioral Ecology and Sociobiology* 55 (2004): 223–230; Denis Boyer, Gabriel Ramos-Fernández, Octavio Miramontes, José L. Mateos, Germinal Cocho, Hernán Larralde, Humberto Ramos, and Fernando Rojas, "Scale-free Foraging by Primates Emerges from Their Interaction with a Complex Environment," *Proceedings of the Royal Society B* 273 (2006): 1743–1750.

The Boston University group's work on the optimality of Lévy flight patterns is published in G. M. Viswanathan, S. V. Buldyrev, S. Havlin, M. G. da Luz, E. Raposo, and H. E. Stanley, "Optimizing the Success of Random Searches," *Nature* 401 (1999): 911–914. See also the minireview G. M. Viswanathan, V. Afanasyev, S. V. Buldyrev, S. Havlin, M. G. E. da Luz, E. P. Raposo, and H. E. Stanley, "Lévy Flights in Random Searches," *Physica A* 282 (2000): 1–12.

On the same subject, see also O. Benichou, M. Koppey, M. Moreau, and R.

Voiturez, "Intermittent Search Strategies: When Losing Time Becomes Efficient," *Europhysics Letters* 75 (2006): 349–354.

For a detailed bibliography of the importance of Lévy flight in subcellular processes, such as the search a transcription factor must perform in order to find the site it needs to attach to on the DNA, see the talks presented at the NORDITA Workshop on Movement and Search: *From Biological Cells to Spider Monkeys,* NORDITA, Stockholm, August 25–27, 2008, available at http://users.physik.tu-muenchen.de/metz/search/search.html. See also Liang Li, Simon F. Nørrelykke, and Edward C. Cox, *"Directed Cell Motion in the Absence of External Signals: A Search Strategy for Eukaryotic Cells,"* PLoS [Public Library of Science] ONE 3 (2008): e2093.

The theorem providing the likelihood that a Lévy particle will return to its release site is called the Sparre Andersen theorem. See Erik Sparre Andersen, *Mathematica Scandinavica 1* (1953): 263; Erik Sparre Andersen, *Mathematica Scandinavica 2* (1954): 195. For a coherent exposition, see also S. Redner, *A Guide to First-Passage Processes* (Cambridge: Cambridge University Press, 2001); J. Klafter, A. Blumen, and M. F. Shlesinger, "Stochastic Pathway to Anomalous Diffusion" *Physical Review A* 35 (1987): 3081.

CHAPTER 18: "Villain!"

The information on Lénárd Barlabási and his descendants comes from the unpublished manuscript of Sándor Barabássy on the history of the Barlabási/Barabási/Barabásy/Barabássy families. Note that for centuries the spelling of the family name was somewhat arbitrary—the same person might use various spellings of his own name interchangeably.

Not all historians accept that the person mentioned in Barlabási's letter is identical to György Székely. The dissenting opinion is discussed in Gábor Barta and Nagy Antal Fekete's *Paraszthaború 1514-ben* (Budapest: Gondolat, 1973), in which they point to a piece of conflicting evidence: While medieval sources refer to György's home village as Dálnok, Barlabási talks about Makfalva in his letter. Given the communal ownership of lands, the movement of men between villages was rather limited. Yet it was not impossible under the *fiú-leány,* or boy-woman, system—a man could move if he married a woman from a different village who was the only heir to her father's land and thus had been treated as a man, her husband inheriting her father's privileges.

CHAPTER 19: The Patterns of Human Mobility

Marta's paper on human mobility was published in M. C. González, C. A. Hidalgo, A.-L. Barabási, "Understanding Individual Human Mobility Patterns," *Na-*

ture 453 (2008): 779–782. For a follow-up paper that offers an example on the importance of understanding human mobility patterns in predicting the spread of mobile viruses, see P. Wang, M. Gonzalez, C. A. Hidalgo, A.-L. Barabási, "Understanding the Spreading Patterns of Mobile Phone Viruses," *Science* 324 (2009): 1071–1076.

The paper in which Sergey, Edwards, and collaborators discuss the challenges of the Albatross data collection is: Andrew M. Edwards, Richard A. Phillips, Nicholas W. Watkins, Mervyn P. Freeman, Eugene J. Murphy, Vsevolod Afanasyev, Sergey V. Buldyrev, M. G. E. da Luz, E. P. Raposo, H. Eugene Stanley, and Gandhimohan M. Viswanathan, "Revisiting Lévy Flight Search Patterns of Wandering Albatrosses, Bumblebees and Deer," *Nature* 449 (2007): 1044–1048.

The Lévy nature of human mobility was discussed widely in the literature following Dirk Brockmann's original proposal. See, for example, I. Rhee, M. Shin, S. Hong, K. Lee, and S. Chong, "On the Lévy-Walk Nature of Human Mobility: Do Humans Walk Like Monkeys?" *INFOCOM 2008. The 27th Conference on Computer Communications. IEEE* April 13–18 2008): 924–932. The concept emerged even in the migrating patterns of tribes—see, for example, C. T. Brown, L. Liebovitch, and R. Glendon, "Lévy Flights in Dope Ju'hoansi Foraging Patterns," *Human Ecology* 35 (2007): 129–138.

The coverage of *Science* magazine on the retraction of the results pertaining to an albatross motion is available at: John Travis, "Do Wandering Albatrosses Care about Math?" *Science* 318 (2007): 742–743. See, however, this later piece, which was written after the Sims data was published, reinforcing the correctness of the Lévy paradigm for many species: Mark Buchanan, "The Mathematical Mirror to Animal Nature," *Nature* 435 (2008): 714–716.

David Sims's article on the foraging patterns in marine animals is available as David W. Sims, Emily J. Southall, Nicolas E. Humphries, Graeme C. Hays, Corey J. A. Bradshaw, Jonathan W. Pitchford, Alex James, Mohammed Z. Ahmed, Andrew S. Brierley, Mark A. Hindell, David Morritt, Michael K. Musyl, David Righton, Emily L. C. Shepard, Victoria J. Wearmouth, Rory P. Wilson, Matthew J. Witt, and Julian D. Metcalfe, "Scaling Laws of Marine Predator Search Behavior," *Nature* 451 (2008): 1098–1102.

CHAPTER 20: **Revolution Now**

For a description of medieval Temesvár and its representation during centuries, see Árpád Jancsó and Loránd Balla, *Temesvár Régi Ábrázolásai, 16–18. század* (Marosvásárhely, Hungary: Mentor Kiadó, 2005).

For a detailed discussion on the ideology of the 1514 uprising and the role of György Dózsa Székely in setting the tone and the course of the events, see Jenő Szűcs, *Dózsa parasztháborújának ideológiája. Nemzet és történelem* (Budapest: 1974), pp. 603–666. See also *Valóság* (1972), vol. 11, pp. 12–39.

A further example of the role of the Szekler tradition in the 1514 revolution is offered by the enduring symbol of the uprising, the bloody stake that the crusaders paraded through each village, calling on the local men to join their ranks. It is rooted in the tradition of the bloody sword carried through Szekler villages to prepare for war. In yet another example, in the proclamation of Cegléd György Dózsa, Székely calls himself a *supremus capitaneus,* a title without precedent in Hungary but one traditionally held by the captain of the Szekler military. Interestingly, in the letter György Székely calls himself a subject of the king, but *"not of the nobility,"* and makes no mention of the Ottomans but calls instead for a war against the *"infidel and mean-spirited nobility."*

For a detailed description of the events surrounding the 1506 Ox Tax, see Lajos Szádeczky Kardoss, *A székely nemzet története és alkotmánya* (Budapest: Franklin Társulat, 1927). See also Ákos Egyed, *A Székelyek rövid története a megtelepedéstől 1918-ig* (Csíkszerda, Hungary: Pallas-Akadémia Kiadó, 2006); István Rugonfalvi Kiss, *A Székely nemzet története* (Máriabesenyő-Gödöllő, Hungary: Attraktor Kiadó, 2004); and Károly Kós, *Transylvania: An Outline of Its Cultural History* (Budapest: Szépirodalmi Könyvkiadó, 1989). For a discussion of the Szekler constitution and rights, see Nathalie Kálnoky, *Les constitutions et privilèges de la Noble Nation Sicule: acculturation et maintien d'un système coutumier dans la Transylvanie médiévale* (Budapest, Paris, Szeged: Publications de l'Institut Hongrois de Paris, 2004). For a Romanian perspective, infused with communist ideology, see Stefan Pascu, *A History of Transylvania,* translated by D. Robert Ladd (Detroit: Wayne State University Press, 1982). For a brief history, see István Lázár, *Transylvania: A Short History* (Budapest: Corvina, 1997).

For a biography of Pál Tomori, who fought the Szeklers over the Ox Tax in 1506, see Vilmos Fraknói, *Egyháznagyok a Magyar középkorból* (Budapest: Élet és Irodalmi Nyomda Részvénytársaság Kiadása, 1913).

One of the places where the link between György Székely and the 1506 battles around the Ox Tax is made is the monumental three-volume *History of Transylvania,* editor-in-chief Béla Köpeczi, edited by László Makkai, András Mócsi, and Zoltán Szász. The 1514 events are discussed in Volume 1: *Erdély Története, A kezdetektől 1606-ig* (Budapest: Akadémiai Kiadó, 1986).

CHAPTER 21: **Predictably Unpredictable**

The Traveler, by John Twelve Hawks, is the first book of the Fourth Realm Trilogy, published by Doubleday Canada (2005).

If you want to gain more insight into Sandy Pentland's work on human dynamics and wearable computing, I recommend his book *Honest Signals: How They Shape Our World* (Boston: MIT Press, 2008).

The work of Nathan Eagle and Sandy Pentland on predicting the location of students is described in several papers. See, e.g., N. Eagle, "Mobile Phones as Social Sensors," *The Handbook of Emergent Technologies in Social Research* (Oxford University Press [in prep.]); N. Eagle and A. Pentland, "Eigenbehaviors: Identifying Structure in Routine," *Behavioral Ecology and Sociobiology* 63, no. 7 (2009): 1057–1066; N. Eagle and A. Pentland, "Reality Mining: Sensing Complex Social Systems," *Personal and Ubiquitous Computing* 10, no. 4 (2006): 255–268.

Our work on the predictability of the mobile-phone users is summarized in Chaoming Song, Zehui Qu, Nicholas Blumm, Albert-László Barabási, *Limits of Predictability in Human Mobility* (submitted for publication).

CHAPTER 22: A Diversion in Transylvania

Kolozsvár, appearing under its Romanian name as Cluj Napoca on most international maps, has a long history dating back to the Romans (Napoca was a Roman colony in the second and third centuries), eventually growing into the capital of Transylvania. For a history of the city, see documents collected in *Kincses Kolozsvár*, vols. I–II, edited by István János Bálint (Budapest: Magvető Könyvkiadó, 1987). See also István János Bálint, *Kolozsvár* (Budapest: Polygon Kiadó, 1989). There is also a rather entertaining picture book designed for children, but informative for adults as well, illustrating life in Kolozsvár from the time of the Romans until today: *Kincses Képeskönyv Kolozsvár* (Kolozsvár, Romania: Koinonia and Projectograph, 2007). See also Lajos Asztalos, *Kolozsvár, Helységnév és településtörténeti adattár* (Kolozsvár, Romania: Kolozsvár Társaság-Polis Könyvkiadó, 2004).

CHAPTER 23: The Truth About LifeLinear

TIA stands for Total Information Awareness, a government program Poindexter oversaw as the director of the Information Awareness Office at the Defense Advanced Technology Agency, or DARPA. Created in the aftermath of the 2001 terrorist attacks, the Information Awareness Office had as its official seal an occult pyramid topped with an all-seeing eye. This Freemason symbol, lifted from the dollar bills fondly stamped by Georgers, vividly captured the agency's goal: to develop technologies to predict and deter future terrorist plots. Poindexter was convinced that the United States could have avoided the 1993 World Trade Center or the 1995 Oklahoma City bombings if software had been available to alert the authorities of suspicious fertilizer purchases by nonfarmers. Then came September 11, 2001, and he was bewildered that the government had failed to prevent this attack as well. So in January 2002 TIA was born. *Scientia est potentia*, or *Knowledge is power*, was the new agency's motto, and Poindexter was determined to prove its worth.

Yet he himself turned out to be a liability for the program. In 1990 he had been convicted for his involvement in the Iran-Contra scandal. The conviction was later thrown out on a technicality, so he never served jail time, but he remained tainted by the process. So when the Orwellian bent of TIA became the magnet of criticism by a wide coalition of privacy advocates and privacy-sensitive lawmakers, Poindexter became their lightning rod. Eventually Congress gave in, and while it did not outlaw the program, it eliminated the funding behind it. A month later, in August 2003, Poindexter resigned from DARPA.

So much has been written about privacy issues in the past decade, particularly given the emergence of Internet-based data-gathering technologies, that it is largely impossible to give justice to the subject. I have a thick folder dealing with everything from privacy in the Facebook and Twitter age to the location-dependent technologies available on smart phones. In the end, given the rapid changes on this topic, I decided against offering a précis as it would be outdated by the time this book goes to print.

The description of the dowry process preceding a wedding is based on Lajos Balázs: *A vágy rítusai—rítusstratégiák. A születés, házasság, halál szokásvilágának lelki hátteréről* (Kolozsvár, Romania: Sciencia Kiadó, 2006), based on two decades of observations in one of the Szekler Land's most traditional villages, Csíkszentdomokos. For an earlier take on the habits, rules, and life of a traditional Szekler village throughout the history, with a series of original documents, see István Imreh, *A rendtartó Székely Falu* (Bucharest: Kriterion Könykiadó, 1973). See also László Barabási, *A Székely Rendtartás* (Budapest: Fríg Kiadó, 2006).

You may also argue, as one of my group members, Max Schich, did upon reading this chapter, that life in a small town is never as idyllic as it looks to an outside observer, and that nonconforming members of a community without privacy must feel trapped in the conventions of the community.

We do not have to go to Transylvania to see Hasan's "privacy-free" philosophy in action once again—take, for example, Timothy Durham, the wrongfully convicted fellow we encountered earlier. "When I was released from prison I had considered developing a device that could be worn just like a pager that could be used to track my movements," he said (see Taryn Simon, Peter Neufeld, and Barry Scheck, *The Innocents* [New York: Umbrage Editions, 2003])—an odd dream for somebody whose life had been monitored twenty-four/seven for the previous six years. As Tim Durham explains, those wearing such a device, if accused of a crime, "could be pinpointed pretty much exactly where they were at the time that crime was committed and prove that they did not commit that crime." Which was exactly Hasan's thought when he developed his tracking project: "What if someone makes a mistake, what if something goes wrong?"

Any discussion about privacy is a discussion about trade-offs. Giving up the privacy of our medical records may allow the insurance companies to refuse coverage, but not sharing the data could limit the quality of the medical care we receive and thwart research toward the development of better cures. Giving up information about our shopping habits may be perceived by some as an uncomfortable loss of privacy, but others are more than willing to part with it for free or discounted services. Giving up information about our employment history and communication patterns may expose us to potential criminal investigation, but may also reward us with higher security and decrease our chances of being caught in a criminal activity or terrorist attack.

Much has been written on the responsibility of scientists, especially given the ethical and philosophical issues surrounding the development of the atomic bomb, as well as those besetting recent explorations into genetics. One of the first and most important treatments of the moral issues that scientists face was the so-called *Franck Report*, or *Report of the Committee on Political and Social Problems* (Manhattan Project "Metallurgical Laboratory," University of Chicago, June 11, 1945), written by James Franck (Chairman), Donald J. Hughes, J. J. Nickson, Eugene Rabinowitch, Glenn T. Seaborg, J. C. Stearns, and Leo Szilárd. See also Soshichi Uchii, "The Responsibility of the Scientist," *Physics Education in University*, 1998, Japan Physical Society (in Japanese, but many translations in English are floating around the World Wide Web).

CHAPTER 24:   Szekler Against Szekler

György Székely's speech before the Temesvár battle, the one we reproduced in this chapter, originally emerged in the chronicle of Giovanni Michele Brutus (Brutus János Mihály, *Magyar Historiája, 1490–1552*). The original was in Latin, but I have used the Hungarian text reproduced in Sándor Márki, *Dósa György* (Budapest: Magyar Történelmi Társulat, 1913), p. 468, as the basis of the quote.

CHAPTER 25:   Feeling Sick Is Not a Priority

For Nicholas Christakis's work on the widower effect, see N. A. Christakis and P. D. Allison, "Mortality after the Hospitalization of a Spouse," *New England Journal of Medicine* 354, no. 7 (2006): 719–730; T. J. Iwashyna and N. A. Christakis, "Marriage, Widowhood, and Health Care Use," *Social Science and Medicine* 57, no. 11 (2003): 2137–2147; and F. Elwert and N. A. Christakis, "Variation in the Effect of Widowhood on Mortality by the Causes of Death of Both Spouses," *American Journal of Public Health* 98, no. 11 (2008): 2092–2098.

Nicholas's work with James Fowler on obesity was published in N. A. Christakis and J. H. Fowler, "The Spread of Obesity in a Large Social Network Over Thirty-two Years," *New England Journal of Medicine* 357, no. 4 (2007): 370–379. I was invited to write the editorial in the journal, placing their findings in the context of network science. See A.-L. Barabási, "Network Medicine—From Obesity to the 'Diseasome'," *New England Journal of Medicine* 357 (2007): 404–407.

Note that Fowler and Christakis have generalized their finding on obesity to a number of other health issues, concluding that happiness (J. H. Fowler and N. A. Christakis, "The Dynamic Spread of Happiness in a Large Social Network: Longitudinal Analysis over 20 Years in the Framingham Heart Study," *British Medical Journal* 337 [2008]: a2338), genetic effects (J. H. Fowler, C. T. Dawes, and N. A. Christakis, "Model of Genetic Variation in Human Social Networks," *PNAS: Proceedings of the National Academy of Sciences* 106, no. 6 [2009]: 1720–1724), and smoking (N. A. Christakis and J. H. Fowler, "The Collective Dynamics of Smoking in a Large Social Network," *New England Journal of Medicine* 358, no. 21 [2008]: 2249–2258) are all affected by social ties. Their findings are summarized in their general audience book, Nicholas A. Christakis and James H. Fowler, *Connected: The Surprising Power of Our Social Networks and How They Shape Our Lives* (New York: Little, Brown and Company, 2009).

The quote on having fewer friends thanks to obesity was left on July 26, 2007, at 8:27 A.M. by a reader who identified himself as jpsmith, as a comment on Gina Kolata's article "Study Says Obesity Can Be Contagious," published in the *New York Times* on July 25, 2007. The article was a synopsis of the results reported by Nicholas Christakis and James Fowler.

The quote from the Szekler's take on death comes from Lajos Balázs: *A vágy rítusai—rítusstratégiák. A születés, házasság, halál szokásvilágának lelki hátteréről* (Kolozsvár, Romania: Sciencia Kiadó, 2006).

Our work on burstiness in disease history is summarized in a working paper, K. I. Goh, N. A. Christakis, A.-L. Barabási, "Scaling and Volatility in Human Healthcare Records" (2009).

Note that one of the earliest indications of burstiness in disease history comes from Darwin himself. The topics of his extensive correspondence varied widely, from the breeding of pigeons to gossip, from love to unrestrained disgust. There was one theme, however, that spanned his letters from youth till death: his health. Darwin loved to write about his changing health, particularly if there was some new malady to report. So much so that he left us 416 letters that describe a wide range of symptoms, from long bouts of vomiting to gut pain, vertigo, dizziness, headaches, cramps and colics, bloating and nocturnal flatulence, severe tiredness, skin problems, including blisters all over his scalp and eczema, crying, anxiety, sensation of impending death, insomnia, and depression. It remains a mystery what caused all the misery Darwin experienced throughout his life. The medical community fails to agree on a definite diagnosis, suspecting everything from psychosomatic disease to panic disorder, Chagas's disease, Ménière's disease, lactose intolerance, Lupus erythematosus, arsenic poisoning, multiple allergies, and hypochondria. There is even a book dedicated to his health travails, Ralph Colp, Jr., *To Be an Invalid, The Illness of Charles Darwin* (Chicago: University of Chicago Press, 1977). The only cure that appears to have worked occasionally was drastic water therapy, consisting of cold showers, vigorous rubbing and body strapping with wet towels, and drinking lots of water. Darwin was so pleased with its beneficial effect that when his daughter, Anne, fell ill, worrying that she had inherited his disease, he took the nine-year-old child to his favorite spa to undergo the treatment that had worked so well for him. She did not survive it.

While Darwin was often ill, he did not always experience pain or discomfort. Indeed, letters written in a few relatively distinct periods carry the vast majority of his complaints, emerging as clear disease bursts in his correspondence pattern. The first is around 1833, followed by a period of good health until 1845, leading to another avalanche of health-related letters around 1849. After about a decade of only sporadic complaints we see a reemergence of his maladies between 1859 and 1863. After 1866 he apparently recovers, the next burst of complaints coming only around 1873, followed by a relatively complaint-free period between 1876 and 1880. Problems reemerge only in 1881, a year before his death. Now, this sequence of bursts is by no means surprising for a person who lived as long as Darwin did—he was seventy-three when he died, at a time when life expectancy in his country was less than forty years. Yet, if one looks at his sequence of health-related letters, it is closer to our e-mail pattern than to a series of random events, suggesting that Darwin's health pattern displays a burstiness similar to the ones we have observed in other human activity patterns.

The study dealing with the impact of Harry Potter on the accidents of children was published in Stephen Gwilym, Dominic P. J. Howard, Nev Davies, and Keith Willett, "Harry Potter Casts a Spell on Accident Prone Children," *British Medical Journal* 331 (2005): 1505–1506. The sequence of events pertaining to the discovery was based on an interview by Amanda Gardner, *Harry Potter Books Keep Kids Safe,* available online on several forums.

The research on the impact of the Red Sox games on emergency-room visits was published in B. Y. Reis, J. S. Brownstein, and K. D. Mandl, "Running outside the Baseline: Impact of the 2004 Major League Baseball Postseason on Emergency Department Use," *Annals of Emergency Medicine* 46, no. 4 (October 2005): 386–387.

The Japanese work on the changes in the activity patterns of depressed individuals was published in T. Nakamura, T. Takumi, A. Takano, N. Aoyagi, K. Yoshiuchi, Z. R. Struzik, and Y. Yamamoto, "Of Mice and Men—Universality and Breakdown of Behavioral Organization," *PLoS* [Public Library of Science] *ONE* 3 (2008): e2050-1-8; T. Nakamura, K. Kiyono, K. Yoshiuchi, R. Nakahara, Z. R. Struzik, and Y. Yamamoto, "Universal Scaling Law in Human Behavioral Organization," *Physical Review Letters* 99 (2007): 138103-1-4.

Ido Golding's paper on burstiness in gene expression is published in I. Golding, J. Paulsson, S. M. Zawilski, E. C. Cox, "Real-time Kinetics of Gene Activity in Individual Bacteria," *Cell* 123, no. 6 (2005): 1025–1036. Yet it is by no means the only publication on burstiness in subcellular processes. See, for example, J. M. Pedraza and J. Paulsson, "Effects of Molecular Memory and Bursting on Fluctuations in Gene Expression," *Science* 319 (2008): 339 and references herein or J. R. Chubb, T. Trček, S. M. Shenoy, and R. H. Singer, "Transcriptional Pulsing of a Developmental Gene," *Current Biology* 16, no. 10 (2006): 1018–1025.

For evidence of discontinuous jumps in the fossil patterns, see the extensive literature on punctuated equilibrium, starting with the original article introducing the concept by Niles Eldredge and S. J. Gould, "Punctuated Equilibria: An Alternative to Phyletic Gradualism," in T. J. M. Schopf, ed., *Models in Paleobiology* (San Francisco: Freeman Cooper, 1972), pp. 82–115, reprinted in N. Eldredge, *Time Frames* (Princeton, N.J.: Princeton University Press, 1985). Gould's books offer a detailed exposition of the empirical data and the evidence supporting their theory: Stephen Jay Gould, *The Structure of Evolutionary Theory* (Cambridge, Mass.: Belknap Harvard, 2002). See also Stephen Jay Gould, *Punctuated Equilibrium* (Cambridge, Mass.: Harvard University Press, 2007), which, however, is fully contained in *The Structure of Evolutionary Theory,* so do not buy both books, as I did.

The concept of burstiness in evolutionary theory was quantified using a series of simple but elegant models introduced originally in Per Bak and Kim Sneppen, "Punctuated Equilibrium and Criticality in a Simple Model of Evolution," *Physical Review Letters* 71 (1993): 4083–4086. For empirical evidence, see Ricard V. Solé, Susanna C. Manrubia, Michael Benton, and Per Bak, "Self-Similarity of Extinction Statistics in the Fossil Record," *Nature* 388, no. 764 (1997).

Per Bak's idea of self-organized criticality, a research area in statistical physics preoccupied with explaining the emergence of power laws in a number of natural systems, was introduced in 1987 in Per Bak, Chao Tang, and Kurt Wiesenfeld, "Self-Organized Criticality: An Explanation of 1/f Noise," *Physical Review Letters* 59 (1987): 381–384. For a general-audience exposition of the ideas behind self-organized criticality and punctuated equilibrium, see, for example, P. Bak, *How Nature Works: The Science of Self-Organized Criticality* (New York: Copernicus, 1996). See also Mark Buchanan, *Ubiquity: The Science of History . . . or Why the World Is Simpler Than We Think* (London: Weidenfeld and Nicolson, 2000).

CHAPTER 26: **The Final Battles**

Once again we rely on Taurinus and other sources to bring closer the two final battles, both of which are surrounded with confusion and contradictory messages. Based on sources quoted by the Barta-Fekete book (p. 199) the Transylvanian forces were believed to have numbered about twenty-two thousand, but that number is probably inflated. The crusaders could have numbered somewhere in the vicinity of thirty thousand, but may have been as few as twenty thousand, given the desertions and the forces that were fighting in other parts of the country, such as Lőrincz's troops at Kolozsvár. The contemporary documents largely agree that the Temesvár battle took place on July 15, 1514. Yet the precise order of events remains a mystery even today. Was there a battle indeed, as most chronicles write about, or were they exaggerated? Was György captured during the battle or before? We will never know. There is also disagreement over the actions of Báthory: The chronicle of Szerémi tells us that Báthory did open the gates and attack the peasants; Istvánffy, however, is convinced of the opposite.

CHAPTER 27: The Third Ear

The Prague meeting I attended was part of Enter3: International Festival of Art, Science, and Technologies, organized by Pavel Smetana in the fall of 2007. The *MutaMorphosis: Challenging Arts and Sciences* conference, where Stelarc and I were the keynote speakers, was held November 8–10, 2007.

For more information on Stelarc and his work, see *Stelarc: The Monograph*, edited by Marquard Smith (Boston: MIT Press, 2005).

A number of recent books, from Ian Ayres's *Super Crunchers: Why Thinking-by-Numbers Is the New Way to Be Smart* (New York: Bantam, 2007) to Stephen Baker's *Numerati* (Boston: Houghton Mifflin Harcourt, 2008) have documented how data mining exploits our deep-rooted predictability.

CHAPTER 28: Flesh and Blood

Probably the best-remembered memento of the György Dózsa Székely historical narrative is his execution. Most people who know of him—and in Hungary and Transylvania the events of 1514 are taught in school—would likely not be familiar with the details of his story, but everybody is familiar with his gruesome end. The same is true for many of his contemporaries—just about anybody who wrote about him discussed his horrific death. Thus several sources offer more or less reliable descriptions of the execution. I have incorporated these sources into this chapter, so the lines in italics come from original sources. For a discussion on the sources that are available on the execution, and their reliability (some of which are believed to have been based on eyewitness accounts), see Barta-Fekete, p. 213.

For more about the work of Tibor Szervátiusz, the sculptor whose work depicting György Dózsa Székely on his burning throne is in the National Gallery in Budapest, see Klára Szervátiusz, *Az idő kapujában: Szervátiusz Tibor szobrairól* (Budapest: Püski, 2003).

A number of novels, dramas, and poems treating the story of György Székely flooded Hungarian literature in the nineteenth and twentieth centuries. The first in the series was Eötvös József, *Magyarország 1514-ben* (Budapest: Magyar Helikon, 1972), published in 1847. Theater works included Sándor Hevesi's *Görögtűz* (Budapest: Atheneum, 1918); György Sárközi's *Dózsa* (Budapest, 1939); Gyula Illyés' *Dózsa György* (Budapest: Szépirodalmi Könyvkiadó, 1956); and the opera by

Ferenc Erkel, Mór Jókai, and Ede Szigligeti, *Dózsa György* (1867). Several famous poems have taken up his cause, some of which, like Endre Ady's *Dózsa György unokája* and Sándor Petőfi's *A nép nevében,* are among the best-known poems in Hungarian literature. For additional historical novels on the topic, see László Geréb, *Égő Világ* (Budapest: Móra Ferenc Könyvkiadó, 1961) and Sándor Gergely, *A nagy tábor, 1514* (Budapest: Atheneum, 1945).

Illustrations

There is a theorem in publishing that each graph halves a book's audience. Its corollary for e-books: Each image halves the number of devices that can properly display it. Join me, nevertheless, in inviting the artist back into the book, making *illuminations*, as the hand-drawn decorations enriching medieval manuscripts are called, relevant once again.

The fifteen images in *Bursts* are the work of the Transylvanian artist Botond Részegh, who left behind many opportunities in two Eastern European capitals so that he could maintain his creative independence in Csikszereda, the small Transylvanian village in the heart of the Szekler Land. He prints his etchings in a tiny windowless closet that his former art teacher allows him to use and relies on a network of friends and fans to get by: Some make the exquisite handmade paper his work is printed on, others maintain his Web site, http://reszeghbotond.ro.

The fifteen illustrations he created for *Bursts* have a mission: to bridge the past and the future, allowing the quintessential historical space and the indispensable facts of science to meet on a single plane. Most images serve two chapters, one focused on history and the other on science. By borrowing elements from both, Botond created a virtual link between them, bringing our visual senses in sync with the text.

Acknowledgments

Science, just like book writing, used to be a lonely process. It has changed significantly in the past decades, however, and today major discoveries increasingly require collaborations and large research teams, rather than lonely scientists. Indeed, as Brian Uzzi and his collaborators, Stefan Wuchty and Benjamin F. Jones, have shown, in the 1950s the highest impact work was published as single-author papers. By 2000, however, much of the extraordinary cited work was authored by groups of individuals with different backgrounds. *Bursts* is a hybrid—while it has a single author, it would not have been possible without the caring help of a large network of colleagues, professionals, friends, and family. It is based on research performed in close collaboration with a number of wonderful and talented students, postdoctoral associates, and colleagues, who have brought their unique skills and perspective to the various projects.

First and foremost, I owe many thanks to those who have joined me in exploring the emerging science of bursts, human dynamics, and mobility, including Eivind Almaas, Nick Blumm, Julian Candia, Zoltán Dezső, Kwang-Il Goh, Marta Gonzales, Cesar Hidalgo, Imre Kondor, András Lukács, Greg Madey, Marcio Menezes, Joao Oliveria, B. Rácz, Chaoming Song, István Szakadát, Zehui Qu, Alexei Vazquez, and Pu Wang.

I am grateful to several individuals who generously shared their stories with me, enriching the narrative: chief among them is Hasan Elahi, whose story has transformed the book; Gary Kenis, who guided me through my first gun show; Dirk Brockmann, who shared his encounters with the dollar bills and introduced me to the Georgers; Sergey Buldyrev, who walked me through his work on albatrosses. I am grateful to Gene Stanley, not only for the support he offered in the past two decades, but also for introducing me to the *refuseniks* and helping me to develop an understanding of the economic processes pertaining to bursts. Jean-Pierre Eckmann was not only willing to recall the genesis of his project on e-mail communication, but also the e-mail data he shared with me was the starting point of my own work on burstiness. I am grateful to Tilman Sauer and Diane Kormos, who compiled the Einstein correspondence record, and Alison M. Pearn from the Darwin Correspondence Project in Cambridge, UK, for sharing with me the time line of Darwin's correspondence. Finally, much of the research described in this book would not have been possible without access to the anonymized mobile phone data set—while its source chose to remain anonymous, I want to thank the many who made this research possible, they know exactly who they are. I am also grateful to Alberto Calero, who has been my mentor when it comes to mobile communications, and whose enthusiasm for the project was and remains contagious.

In writing the historical narrative the help of Erdős Heidi, the expert librarian at Collegium Budapest, was indispensable—she not only tracked down the 1507 letter, prompting my journey to Nagyszeben, but also located many of the key monographs, chronicles, and documents that were essential for my understanding of the 1514 events. I am also grateful to József Darvas Kozma of Csíkszereda, who provided the first rough translation of the 1507 letter and helped me find documents pertinent to the era, and thanks to Daniel Perett for the careful English translation. Much of the information about Lénárd and his descendants came from Sándor Barabássy, whose passion for the era is absorbing. I am also thankful for Miklós Köllő for helping me to understand the fighting techniques of the sixteenth century Szeklers. I cannot forget the ladies at the Saxon Archives in Nagyszeben, whose names I never learned, yet who put aside protocol and rule, allowing me a peek at the medieval letter. Many thanks to Gábor Kolumbán, the consummate Transylvanian hub, who activated his numerous links around the region, smoothing my path to the archives. Finally, I am thankful for Lawrence Cunningham for his guidance on the medieval Mass.

Several individuals have worked with me during the past four years—without their dedicated editing I would have never been able to finalize the book. Enikő Jankó and Ágnes Petróczky patiently helped me through the more than thirty hand-corrected versions of each chapter. Suzanne Aleva, Trevor Gillaspy, Nicole Haley, Nicole Leete, and Alina Mak stepped in during the various stages of this project, whenever there was a jam in the pipeline. In addition, Nicole Haley collected the press related to Tim Durham, and Nicole Leete did some great editing in the early stages of the manuscript. The manuscript has greatly benefited from the exceptional editing skills of Deborah Justice, who carried over her gentle touch from *Linked* to *Bursts*. Many thanks to Suzanne Aleva, who fiercely and successfully guarded my seclusion during the lengthy writing process.

Several friends and colleagues were kind enough to read the manuscript in its various stages and offered formidable comments. I have benefited from the careful reading of Dániel Barabási, József Baranyi, János Kertész, Maximillian Schich, and Eduaro Zambrano—each of them brought a valuable point of view to the project. In most cases I accepted their suggestions and proposals with gratitude.

I have benefited from insightful discussions with Kristóf Nyíri on the nature of time and its philosophical implications and Imre Kondor, who provided me the first data showing the bursty nature of economic processes and helped me spend some time at the Reifessein Bank's trading desk in Budapest, trying to understand the nature of bursts in currency trading. I have also benefited from extensive communications with Frank R. Baumgartner and Bryan D. Jones on the intricacies of decision making in legislature and government, and Brian Uzzi gave me a series of leads on the potential role of bursts in innovations. It is a pity that the chapter on the economy had to be dropped at the very end.

The illustrations in the book were the result of a yearlong collaboration with the talented young artist Botond Részegh, who worked tirelessly until we found just the right harmony between science, art, and history. We were also helped with images and technical assistance from Csongor Bartus, Robert Ördög volunteered his extensive collection of old postcards, Zoltán and Zsuzsa Néda on short notice provided us with the photos of the St. György sculpture in Kolozsvár, and Lajos Balázs gave us the image of the dowry women in Csikszentdomokos. Nina Sellars graciously allowed us to use her photograph "Stelark: Extra Ear on Arm" in the image preceeding Chapter twenty-six. Last, I am thankful for Sándor Kányádi, a treasure of the Szekler Land and Transylvania, for introducing me to Botond and thus initiating this wonderful collaboration.

Finally, I am grateful for Katinka Matson and the staff of Brockman Inc. for finding a home for the book and for Stephen Morrow of Dutton, who not only believed in the project from day one but has patiently nurtured it.

These days a book does not end with writing it—I have had lots of fun working on the online incarnation of *Bursts* with Nick Blumm, Johanne Fantini, Isabel Meirelles, and Alec Pawling: our "secret project," as we have come to call it. If you have not yet seen the product of their creative work, go to http://brsts.com.

Then there is the power of the place—did you ever think *where* is a book written? Offices, attics, kitchen tables? The question is not irrelevant—the space gets imprinted into the book's DNA. I have moved several times while writing *Bursts,* so I have worked on it from Back Bay to South Bend, from Csikszereda to the Buda Castle, and from Newton to Brookline. It was finalized in cafeterias—Café Fix and Athans in Brookline, numerous Starbucks along my path, and the Lincoln Street Coffee House in Newton Highlands. At the end I developed a deep appreciation for these urban patrons of creative work, helping legions of writers turn coffee into books.

A lot happened during the four years I worked on this book: I was blessed with two wonderful children, I moved four times, changed jobs, and in the final miles of the project I broke my wrists in a bike accident— each of them good enough reasons to give up and refocus on life. At the end I kept going—a perseverance that would not have been possible without the understanding of my partner and wife, Janet. I will always be profoundly grateful for her.

Index

AdSense program, 222
Afanasyev, Vsevolod, 158, 173
age and predictability, 203
albatross flight patterns, 158, 160–61, 173–74, 174–75, 179
Albert Einstein Archives, 135–36
alchemy, 47
Ali of Epirus, 37–42, 73, 74, 101
Almaas, Eivind, 46
Amazon.com, 255
animal travel, 160, 161–62, 175, 179
antiterrorism, 11–12
Apátfalva conflict, 91–96, 107, 110, 130
aquatic species, motion of, 178–79
Ayres, Ian, 255

Bak, Per, 241n
Baker, Stephen, 255
Bakócz (cardinal)
    appointment of Székely, 74–76
    canceling of Crusade, 129, 131
    and Hungarian independence, 259
    initiation of Crusade, 55–60, 74–76, 113
    and leadership of Crusade, 125–26
    and papal election, 15–19, 56, 68
    and peasant uprising, 144–45
    recruitment for Crusade, 92–93, 94, 113
    and Telegdi's warning, 258
Bánffy, János, 245, 246, 260n
banks, 255
Barlabási, János, 210n
Barlabási, Lénárd

and Kolozsvár battle, 212, 228, 244–45, 259
and Székely's pre-Crusade history, 165–67, 170, 186, 188
Bartholinus, 243
Báthory, István
    and Apátfalva conflict, 95, 109
    defeat of, 210
    and leadership of Crusade, 76
    and Nagylak battle, 110–11, 112, 114
    and Temesvár siege, 183, 188–89
Bebek, János, 185
Becquerel, Henri, 48
believers, 64
Bell, Gordon, 9–10
Beriszló, Péter, 95
birds, 157, 158. See also albatross flight patterns
Blumm, Nick, 198
Boltzmann, Ludwig, 195n
Boston Red Sox, 236–37
Branderburg, György, 210
British Atlantic Survey, 174
British Meteorological Service, 62
Brockmann, Dirk
    and eye movements, 157
    and human mobility, 50, 176
    and Lévy patterns, 157, 162–63, 178, 179
    and mobile communication patterns, 172
    and movement of money, 23–25, 30, 31, 32, 63, 68, 155–57, 162–63
    and Nature article, 155–57
    super-diffusive law, 31

Brown, Robert, 27
Bruto, Gianmichele, 260–61
Brutus, 243
Buda fortress, 185
Buldyrev, Sergey, 158–61, 173–74, 175, 178, 179

Callixtus III, Pope, 38
Capistrano, Giovanni da, 38–39, 71
Castellesi (cardinal), 56
casualties, 101–2
cavalries, 109–15, 226–27
cell biology, 162, 238–41
chaos theory, 67
Chicago Jury Project, 81, 83
Christakis, Nicholas, 229–32, 233
Christianity, 113, 278
chroniclers, 260–61
Collegium Budapest, 44–45
computers, 193
Constantinople, 57, 75, 91, 125–26. *See also* Crusade
Cook, Marshall J., 119
Council of Pisa, 125
Crusade
   canceling of, 130, 144
   compulsory support of, 113–14
   initiation of, 55–60, 68
   leadership of, 72–76, 126 (*see also* Székely, György)
   and papal politics, 126
   recruitment for, 92–93, 94, 110, 113, 210n
   and Telegdi's warning, 55, 56–57, 58, 59–60, 63, 64, 259–60
   See also crusaders
crusaders
   Apátfalva conflict, 91–96, 107, 110, 130
   and capture of leaders, 247–48
   and chroniclers, 260–61
   clemency offers, 244
   end of peasant uprising, 247–48
   Kolozsvár battle, 207–12, 227–28, 244–45, 246–47
   Nagylak battle, 109–15, 130, 184, 260n
   Temesvár battle, 225–27, 228, 243–44, 245–46, 247–49
   Temesvár siege, 183–84, 185–86, 188–89, 210
   threats against, 211
   See also serfs and peasants
Csáky, Miklós

Apátfalva conflict, 95
   execution of, 144, 168
   Nagylak battle, 111, 112, 114
   and recruitment for Crusade, 110
   and Székely's pre-Crusade history, 74, 145, 168, 188
   and Székely's reward, 168, 205
Csanád, 185, 210
Curie, Pierre, 48

Darwin, Charles, 137, 140, 239–40, 291–92
Darwin Correspondence Project, 137
data mining, 256, 256n
Dávid, Francis, 258–59
depression, 237–38
Derryberry, Dennis, 24, 30, 155–56, 272
Dezső, Zoltán, 46
diffusion theory
   Einstein's development of, 26–27, 29, 33
   and human mobility, 29–30
   and Lévy patterns, 157
   and movement of money, 31, 33, 51, 63–64, 68, 156–57
   and Poisson distributions, 84
disease, 24–25, 28–29, 32, 232–34
DNA and transcription factors, 162
dowries, 216–18
Drágffy, János, 245, 246
Durham, James, 78
Durham, Tim, 78–79, 80–81, 83, 86–87, 289

Eagle, Nathan, 193–94
ear, prosthetic, 251–53
Eckmann, Jean-Pierre, 99–100, 103, 281
Eddington, Arthur Stanley, 142
Edison, Thomas, 117
Edwards, Andrew, 174–75, 178
efficiency, 117–18
Einstein, Albert
   correspondence of, 135–38, 139, 140, 140n
   diffusion theory, 26–27, 29, 31, 33, 51, 68, 84, 156, 157
   general relativity theory, 141–42
   and Kaluza, 133–35, 136, 138–39, 140–41, 142–43
   and quantum theory, 66, 67
Einstein Papers Project, 136
Elahi, Hasan
   detentions of, 1–7, 219–20
   extralegal status of, 269
   and FBI, 5–9, 11–12, 79, 268–70

mobility patterns of, 177
outlier status of, 11–12, 270
and photography, 104–5
and polygraph test, 268–69
privacy of, 221–22
on rationality, 79
self-surveillance of, 8–9
unpredictability of, 195, 200–202
Eldredge, Niles, 240
e-mail
    and letter-based correspondence, 138
    patterns in, 98–99, 100–101
    and power laws, 103
    and queueing, 280
    and Web-browsing habits, 104
emergency room visits
    and Boston Red Sox, 236–37
    and Harry Potter, 234–36
English language, 196n
entropy
    defined, 195n
    and English language, 196n
    and knowledge of past movements, 204
    and predictability, 195–97, 198–99, 205, 256
Eötvös, József, 167
Erdős, Hédi, 149
Europeans, 220
evolution, 239–41
eyes movements, 157

fame, fifteen minutes of, 43, 44, 49
Federal Bureau of Investigation (FBI)
    and Hasan Elahi, 5 –9, 11–12, 79, 268–70
    searches for evidence of, 161
fifteen minutes of fame, 43, 44, 49
foraging patterns, 160, 161–62, 175, 179
Fowler, James, 230–32
Framingham Heart Study, 230–32
Franciscan monks and friars, 92, 113, 186, 281
free will, 105, 254
friends, 218, 229–32

Garrison, Jess, 86–87
Gaussian distribution, 157n, 178, 199
Geisel, Theo, 30, 157
Gelfand, Alan E., 83
gender and predictability, 203
Goh, Kwang-Il, 233
gold, 51

Golding, Ido, 239
González, Marta, 172, 176, 195
Google, 11, 45, 220, 222
Gould, Stephen Jay, 240
government, 220
Griessman, Eugene, 119
Gwilym, Stephen, 235
Gyulafehérvár, bishop of, 210n

Hamdan, Salim, 87
Harder, Uli, 104
Harlequins of *The Traveler*
    and Hasan Elahi, 202
    and random-number generators, 192, 194, 203
    unpredictability of, 195, 195n
    and the Vast Machine, 191
Harry Potter series, 234–36
Havlin, Shlomo, 161
health, 229–41
    and cell biology, 238–41
    depression, 237–38
    and emergency room visits, 234–37
    and obesity, 231–32
    and priorities, 232–34
    and relationships, 229–32
Hebrich, Conrad, 26, 66
Heisenberg, Werner, 197
Hesburgh Library at the University of Notre Dame, 104
Hidalgo, Cesar, 176
historicist doctrine, 65
history, 65–66
History of Transylvania, 188
history's role in predictability, 204, 205
Homeland Security, 12, 200–201, 202, 219–20, 269
home sales, 64
hubs of networks, 119
Hufnagel, Lars, 30
human dynamics, 256n
human mobility, 171–80
    and diffusion theory, 29–30
    and dollars (*see* money, movement of)
    and entropy, 198–99
    and Lévy patterns, 177, 179–80
    MIT study on, 193–95
    mobile-phone data, 171–72, 175–78, 179–80, 195
    and power law, 177n
Hungarian Historical Society, 148–49, 168

Hungarian National Gallery, 267
Hungary
 defenses of, 144, 183, 208
 and fortress at Belgrade, 38
 and György Székely, 144, 147, 184, 185, 223
 independence of, 259
 internet traffic in, 45–46, 49
 king of, 58–59, 185, 187, 209, 248, 258
 population explosion in, 58
 serfs of, 58
 See also specific locations and leaders
Hunyadi, János, 38–39, 58, 71–72, 92–93, 183–84

Innocence Project, 86
Institute for Advanced Studies, Budapest, 44–45, 149
insurance, 237, 255
intelligence agencies, 219
interdependence and privacy, 218
Internet documents, life spans of, 105–6
Iraq invasion (2003), 277
Istvánffy, 148, 243, 247

Jankó, Eniko, 218
Jewish National Library, 135
John I, King of Hungary, 258
John II, King of Hungary, 258
John Radcliffe Hospital, Oxford, England, 234
Jovius, 243
Julius II, Pope, 56, 125
juries, 79, 80–81, 83, 84, 87

Kaluza, Theodor
 and Einstein, 133–35, 136, 138–39, 140–41, 142–43
 multidimensional universe theory, 133–34, 136, 143–44
Kanis, Gary, 21–23, 26, 30
Karadzic, Radovan, 214
knowledge, 241
Kolozsvár conflict, 207–12, 227–28, 244–45, 246–47
Kormos-Buchwald, Diana, 135–36
László-Herbert, Mark, 150
Lázár, András, 170, 212, 245
leadership, 259
Lee, Ivy, 118, 122, 143
Leo X, Pope
 and Bakócz, 125–26

 and Crusade, 56, 57
 election of, 55
letter-based correspondence, 135–38, 139–40, 140n, 142
Lévy, Paul, 157n
Lévy patterns
 about, 157
 in albatross flight patterns, 160, 173–74, 179
 in animal travel, 157, 161, 175, 179, 284
 in eye movements, 157
 and György Székely's path, 163, 180
 in human mobility, 177, 179–80
 in mobile communication patterns, 172
 in movement of dollars, 157, 162–63, 178, 179
 in particles, 157, 163
 and power law, 157n
 in scientific progress, 179
LifeLinear surveillance enterprise, 213–16, 218–20, 222
Linked (Barabási), 45
Lorentz, Hendrik Antoon, 142
Lőrincz of Bihar
 capture evaded by, 249
 and chroniclers, 261
 and Kolozsvár battle, 208–12, 227–28, 245, 246
 and Nagyvárad, 185
 and Telegdi, 260n
Lőrincz of Cegléd, 243, 259
Lukács, András, 45–46

Mach, Ernst, 27
The Magic Number Seven (Miller), 140n
magnetic dipoles, 66
Mandelbrot, Benoît, 157n
Marine Biological Association Laboratory, 178
Márki, Sándor, 168
Massachusetts Institute of Technology (MIT), 193–95
Mateos, José Luís, 284
mathematical models, 63–64
Mathias Church, 71–72, 275
Matthias I, King of Hungary, 208
Maxwell, James Clark, 134
Medicare, 233
Medici, Giovanni de', 16, 17, 55, 56. *See also* Leo X, Pope
Mehmed II, 38, 39, 58

Miller, George, 140n
mobile-phone data, 84–85, 171–72, 175–79, 195, 205
Molly (rape victim), 77–78, 79–80
money, movement of
  and diffusion theory, 31, 33, 51, 63–64, 68, 156–57
  and human mobility, 26, 30, 175–76
  and Lévy patterns, 157, 162–63, 178, 179
  Nature article on, 32, 155–56
  and Poisson process, 106
  See also WheresGeorge.com
monkeys, 157, 284

Nagyenyed, 211, 212
Nagylak battle, 109–15, 130, 184, 260n
Nagyszeben, 147–54
Nagyvárad fortress, 185
National Association of Realtors, 64
National Institutes of Health, 230
Nature, 32, 155–56, 160, 173, 175, 178
Netflix, 255
network hubs, 119
news, life span of, 44, 46, 49–50, 63–64, 105
Newton, Sir Isaac, 48
New Yorker, 125
Nostradamus, 64
Numerati (Baker), 255

obesity, 231–32
Odeporicon, 148
Oliveira, João Gama, 137
omnipresence of bursts, 120–21
Origo.hu, 45–46, 49–50
Oswald, Wilhelm, 27
outliers
  and Gaussian distribution, 178, 199
  György Székely, 12
  Hasan Elahi, 11–12, 270
  and human mobility, 177–78, 195
  implications of, 270
  and movement of money, 178
  and Poisson distribution, 102, 178
  and power laws, 102, 178, 199, 202
Paczusky, Maya, 104
pandemics, 24–25
papal election, 15–19
Pareto, Vilfredo, 102
particles
  and Lévy patterns, 157, 163

and the uncertainty principle, 197
peasants. *See* serfs and peasants
Pentland, Sandy, 193
Perényi, István, 185
Perett, Daniel Gregory, 166
perne traditions, 216–18
Perrin, Jean-Baptiste, 28, 30, 33
Petrovics, Peter, 247, 248
Phillips, Richard, 174–75
Philosophiae Naturalis Principia Mathematica (Newton), 48
photographs, 104–5
Poindexter, John, 215, 288–89
Poisson, Siméon-Denis
  prioritizing of, 123
  scope of research, 82
Poisson distributions
  described, 84–86
  and e-mail, 98–99, 100
  and jury reliability, 82–84, 87–88
  and movement of money, 106
  and outliers, 102, 178
  and power laws, 120, 123
  and predictability, 83–84
  and RNA-molecule generation, 239
pollen, 27–28
Pool, Peter, 159
Popper, Karl, 65–66, 255, 256–57, 260
power laws
  and albatross flight patterns, 160
  and correspondence, 103, 137–42
  and depression, 238
  described, 102–5
  exponents of, 138, 138n
  and human mobility, 119–20, 172, 177, 177n, 178, 198–99, 202
  and Lévy patterns, 157, 157n
  and Mathematica programming, 123–24
  and movement of money, 106, 157
  and outliers, 102, 178, 198–99, 202
  pervasiveness of, 105, 119
  and Poisson distributions, 120, 123
  and predictability, 199, 202
  and priorities, 124–25
power laws (cont.)
  and procrastination, 281
  and Richardson's law, 277
  and Web-browsing habits, 104
predictability, 61–68
  and commercial interests, 257
  and data mining, 256

predictability (*cont.*)
  and entropy, 195–97, 198–99, 205
  and Gaussian distributions, 157n, 178,
    199
  history's role in, 204, 205
  at individual level, 33, 255
  limits of, 197–98
  MIT study on, 193–95
  and Poisson distributions, 83–84
  Popper on, 255, 256–57
  and power laws, 202
  and redundancy, 196–97
  and the uncertainty principle, 197
  and youth, 193
"Prediction and Prophecy" (Popper), 65
priorities
  and burstiness, 121–22, 124
  and consequences, 143
  and disease occurrence, 234
  effectiveness of, 125
  and Einstein's correspondence, 139
  and health issues, 234–37
  lists of, 118–19
  number of tasks on, 140n
privacy
  of Hasan Elahi, 221–22
  and interdependence, 218
  prospective privacy, 223
  Stelarc on, 253
  and surveillance, 10, 216, 219, 253–54
  universal sense of, 220
Proclamation of Cegléd, 186
prophecy, 65
prospective privacy, 223
Protestant Reformation, 258
punctuated equilibrium, 240–41
purchase preferences, 255

Qu, Zehui, 194, 201, 201n
quantum theory, 66–67
queueing, 124–25, 280

radium atoms, 48–49
Radoszlav, 185, 244, 248
random walks, 157n, 158, 161, 163
redundancy, 196–97, 196n
relationships, 218, 229–32
repetitive systems, 65, 255
research on human behavior, 220–21, 222
revisionists, 100
Riccardus, Bartholi, 148

Richardson, Lewis Fry
  on conflict prediction, 97–98, 101, 106–7,
    113, 258, 277
  failure of publications, 257–58
  on weather prediction, 61–63, 66, 86,
    257–58
RNA-molecule generation, 239
Romanian military units, 227, 228, 245
rovás script, 148–49
Roy, Deb, 10
Ruth, Babe, 236
Rutherford, Ernest, 48, 50

Sauer, Tilman, 136, 137
Saxon National Archives in Nagyszeben,
    149, 150–51, 152–54
Saxons
  and András Lázár, 212
  and Kolozsvár battle, 245–46
  and Szeklers, 187–88
  and Temesvár battle, 226, 227, 228
Schwab, Charles Michael, 117–18
science, 179, 257
Science magazine, 175
Seaborg, Glenn, 51
self-organized criticality, 241n
serfs and peasants
  and Apátfalva conflict, 91–96, 101, 109–10
  and Bakócz's call to arms, 58
  and canceling of Crusade, 130
  and capture of leaders, 248
  and class distrust, 258
  and clemency offers, 244
  and execution of Székely, 266
  and Hunyadi's mission, 39, 58
  and Kolozsvár conflict, 245, 246–47
  and land of aristocracy, 186
  and landowners, 58, 93–94, 113, 114, 147
  and Nagylak battle, 111, 112, 113, 184
  and recruitment for Crusade, 110, 210n
  regard for Székely, 76
  and taxes, 58
  and Telegdi's warning, 58, 59, 60, 63,
    259–60
  and Temesvár battle, 226, 227, 248
  and Temesvár siege, 184, 189
  and uprising, 114, 129–32, 144, 186, 248
  See also crusaders
Shannon, Claude, 196n
Sheck, Barry, 86
shopping preferences, 255

shrouding element of bursts, 196
similarities between people, 254
Simó, György, 45
Sims, David, 178, 179
skeptics, 64
smart phones, 193–94
Smith, Clive Stafford, 201
snapshots, 104–5
social sciences, 65–66
Soddy, Frederick, 48
Solomon, Herbert, 83
Song, Chaoming, 195–97, 198
Stanley, Gene, 159–60
The Statistics of Deadly Quarrels
    (Richardson), 97–98, 101, 258
Stelarc, 252–53, 256
Stierochsel, István, 210n. *See also* Taurinus
string theory, 144
suicide, 106, 237
Super Crunchers (Ayres), 255
super-diffusive law, 31
surveillance
    and Hasan Elahi, 8–10
    LifeLinear surveillance enterprise, 213–
        16, 218–20, 222
    and research on human behavior, 220–21
    Stelarc on, 253
Szabó, Károly, 149, 160
Szapolyai, János
    amassing of forces, 225–28
    and Apátfalva conflict, 95
    and Bakócz, 59, 76
    as king of Hungary, 258
    and Kolozsvár battle, 209–12
    and leadership of Crusade, 76
    opposition to Crusade, 59, 209–10
    and recruitment for Crusade, 210n
    and Temesvár battle, 243–44, 245–46,
        247
Szárhegy, 169–70
Székely, Gergely
    and Ali of Epirus, 39–41, 273
    capture of, 247, 249
    and Crusades, 73
    execution of, 264, 265
    and Szekler background, 186
    and Temesvár battle, 243
    and uprising, 185
Székely, György
    and Ali of Epirus, 39–42, 68, 73
    and Apátfalva conflict, 94–96, 107, 109

appointed leader of Crusade, 72–76, 126
    authority of, 204, 254n
    burst of, 254
    canceling of Crusade, 129–32
    capture of, 247, 249
    and chroniclers, 260–61
    criminal accusations, 166–68, 188
    execution of, 263–67, 294
    history's account of, 168, 185–86, 267–68
    and king, 254n
    leadership of, 185–87
    and Lévy patterns, 163, 180
    and Nagylak battle, 111–12, 114
    name of, 12, 145, 148
    outlier status of, 12
    popular understanding of, 147–48
    pre-Crusade life of, 145, 148, 149, 166–
        70, 186–88
    priorities of, 115
    recruitment and training of volunteers,
        92–94
    reward promised to, 74, 168, 205
    sculpture of, 267
    and Szekler background, 186
    and Temesvár battle, 225–27, 243–44, 247
    and Temesvár siege, 183–84, 188–89
    uprising, 130–32, 144–45
Szeklers
    history of, 148–49
    and King Ulászló, 209
    and Kolozsvár battle, 245–46
    military units, 226–27, 228
    and political upheaval, 187–88
    status of, 114
    and Székely's pre-Crusade history, 148,
        166–67
    and Temesvár battle, 247
    traditions and culture of, 186, 248
Szerémi, 148, 260n, 261
Szolnok, 186

Tang, Chao, 241n
Taurinus
    acknowledgement of benefactor, 156
    on Apátfalva conflict, 95
    and bishop of Gyulafehérvár, 210n
    on Kolozsvár battle, 244–45, 246–47
    on Nagylak battle, 111, 112, 113
    on pre-Crusade life of Székely, 148
    on Temesvár battle, 243, 247
technological advances, 7

Telegdi, István
   and Bakócz, 258
   and chroniclers, 259–61
   death of, 246, 260n
   opposition to Crusade, 58, 261
   predictions of, 55, 56–57, 59–60, 63, 64,
      68, 84, 107, 145, 255, 261
   and Székely, 74, 168
Telegdi, János, 76, 188, 268
Temesvár fortress
   final battle for, 225–27, 228, 243–44,
      245–46, 247–49
   initial siege, 183–84, 185–86, 188–89, 210
Theory of Everything, 134, 143
Thornallai, János, 245, 246
Time, 43
time-management, 119
Time Management (Cook), 119
Time Tactics of Very Successful People
   (Griessman), 119
Tokyo University, 237–38
Tomori, Pál, 187
Total Information Awareness (TIA)
   program, 215–16, 219, 288
Tracking Transience website, 9, 221–22
transcription factors, 162
transmutation, 47–48, 51, 68
transportation, 29
Transylvania
   and Crusade, 93, 94, 95, 109, 209–10, 210n
   defenses of, 226, 228, 244, 247, 248
   and György Székely, 114, 147, 148, 163,
      166–67
   religious tolerance in, 258
   as semi-independent principality, 258
   and Szeklers, 12, 39, 148, 149
   three nations in, 187–88
   wedding traditions in, 216–18
   See also specific locations and leaders
The Traveler (Twelve Hawks), 191–92, 194
Tubero, 148, 260
Twain, Mark, 12–13

Twelve Hawks, John, 191–92
Ulászló
   and Bakócz, 58–59
   and capture of Székely, 248
   and Jagellion dynasty, 274
   and nationality of successor, 209
   and Szapolyai, 59
   and taxes, 187
   and uprising, 185
uncertainty principle, 197
Unitarian church, 258–59
The United States Currency-Tracking
   Project. *See* WheresGeorge.com
Unus, Jehova Sanctus, 47–48
U.S. Supreme Court, 87

Vast Machine, 191, 192, 215–16, 219
violence, 77–88
Viswanathan, Gandhi, 160, 161
Vlad the Impaler, 76

Warhol, Andy, 43, 44, 49
wars, laws governing, 97–98, 101, 106–7
wearable computing, 193
weather prediction, 61–63, 66, 86
website visitation, 50, 68, 104
wedding traditions in Transylvania, 216–18
WheresGeorge.com
   about, 25–26
   and Derryberry, 30, 155–56
   and diffusion of dollars, 30–32
   and human mobility, 175–76
   Kanis's introduction to, 22–23
   and power laws, 106
   See also money, movement of
widower effect, 229
Wiesenfeld, Kurt, 241n
wrongful convictions, 81, 83, 87

Yahoo!, 11
youth, 203